Benefit-sharing in Environmental Governance

Taking a bottom-up perspective, this book explores local framings of a wide range of issues related to benefit-sharing, a growing concept in global environmental governance.

Benefit-sharing in Environmental Governance draws on original case studies from South Africa, Namibia, Greece, Argentina, and Malaysia to shed light on what benefit-sharing looks like from the local viewpoint. These local-level case studies move away from the idea of benefit-sharing as defined by a single international organization or treaty. Rather, they reflect different situations where benefit-sharing has been considered, including agriculture, access to land and plants, wildlife management, and extractives industries. Common themes in the experiences of local communities form the basis for an exploration of spaces for local voices at the international level in the Convention on Biological Diversity (CBD), often argued to be the most open arena to non-state actors, and therefore vital to how local voices may be included at the global level. The book analyzes the decisions of the CBD parties to produce an in-depth reflection on how this arena builds and delimits spaces for the expression of local community themes, and paths for local community participation including community protocols. The book then situates the bottom-up findings in the wider debate about global civil society and deliberative democracy in environmental governance.

This interdisciplinary book will be of great interest to students and scholars of environmental politics, environmental law, political ecology and global governance, as well as practitioners and policymakers involved in multilateral environmental agreements.

Louisa Parks is Associate Professor of Political Sociology at the School of International Studies and the Department of Sociology and Social Research, Università degli studi di Trento, Italy. She was a research fellow for the BeneLex project on fair and equitable benefit-sharing in global environmental law, and has published work on community protocols and local voices in the Convention on Biological Diversity, as well as civil society, social movements, and the impacts of their campaigns.

Earthscan Studies in Natural Resource Management

Sustainability Certification Schemes in the Agricultural and Natural Resource Sectors
Outcomes for Society and the Environment
Edited by Melissa Vogt

Policy and Governance in the Water-Energy-Food Nexus
A Relational Equity Approach
Edited by Anastasia Koulouri and Nikolai Mouraviev

UNESCO Biosphere Reserves
Supporting Biocultural Diversity, Sustainability and Society
Edited by Maureen G. Reed and Martin F. Price

Social Movements Contesting Natural Resource Development
Edited by John F. Devlin

Benefit-sharing in Environmental Governance
Local Experiences of a Global Concept
Louisa Parks

Governing Renewable Natural Resources
Theories and Frameworks
Edited by Fiona Nunan

Interdisciplinary Collaboration for Water Diplomacy
A Principled and Pragmatic Approach
Edited by Shafiqul Islam and Kevin M. Smith

For more information on books in the Earthscan Studies in Natural Resource Management series, please visit the series page on the Routledge website: http://www.routledge.com/books/series/ECNRM/

Benefit-sharing in Environmental Governance

Local Experiences of a Global Concept

Louisa Parks

Routledge
Taylor & Francis Group
LONDON AND NEW YORK

earthscan
from Routledge

First published 2020
by Routledge
2 Park Square, Milton Park, Abingdon, Oxon OX14 4RN

and by Routledge
52 Vanderbilt Avenue, New York, NY 10017

Routledge is an imprint of the Taylor & Francis Group, an informa business

First issued in paperback 2021

British Library Cataloguing-in-Publication Data
A catalogue record for this book is available from the British Library

Library of Congress Cataloging-in-Publication Data
Names: Parks, Louisa, 1979- author.
Title: Benefit-sharing in environmental governance : local experiences of a global concept / Louisa Parks.
Description: Abingdon, Oxon ; New York, NY : Routledge, 2020. | Series: Earthscan studies in natural resource management | Includes bibliographical references and index.
Identifiers: LCCN 2019032314 (print) | LCCN 2019032315 (ebook) | ISBN 9780367181871 (hardback) | ISBN 9780429198311 (ebook)
Subjects: LCSH: Environmental management—Case studies. | Environmental justice—Case studies. | Environmental policy—Case studies. | Sustainability—Case studies.
Classification: LCC GE300 .P36 2020 (print) | LCC GE300 (ebook) | DDC 333.7—dc23
LC record available at https://lccn.loc.gov/2019032314
LC ebook record available at https://lccn.loc.gov/2019032315

ISBN: 978-0-367-18187-1 (hbk)
ISBN: 978-1-03-208391-9 (pbk)
ISBN: 978-0-429-19831-1 (ebk)

Typeset in Sabon
by codeMantra

Visit the eResources: www.routledge.com/9780367181871

Contents

Acknowledgements

First and foremost, my sincere thanks to all of the local community members who were so welcoming and patient in their explanations, and kind enough to take part in the research. This book does not claim to understand or represent the realities of your experiences, but any element that does ring true is thanks to your involvement. My thanks also go to all of those that participated in interviews for this research. The local community case studies were carried out alongside NGO partners and a host of others. Heartfelt thanks to Maria Gaglia-Bareli and Frosini Koutsouti of the Ikarian Documentation and Research Center (Greece); the late Rodney Sibuye of the Kukula Association of Traditional Health Practitioners of Bushbuckridge, Cath Traynor and Gino Cocchiaro at Natural Justice, Marie-Tinka Uys at the Kruger to Canyons Biosphere Reserve, and Prof. Wayne Twine of the University of the Witwatersrand Rural Facility (South Africa); Lesle Jansen at Natural Justice and Friedrich Alpers at Integrated Rural Development and Nature Conservation (Namibia); Ana di Pangracio and Pia Marchegiani at the Foundation for the Environment and Natural Resources (Argentina); Prof. Poline Bala of the University of Sarawak and – last but definitely not least – Harry Jonas at Natural Justice (Malaysia), all of whom helped a great deal to improve previous drafts of the work contained in this book. Research trips would have been much the poorer without your wisdom, encouragement, dedication, and good humour. Harry, your energy and enthusiasm about what this work could be was inspiring. I hope it comes somewhere remotely close.

The research for this book was carried out as part of the BeneLex project (Benefit-sharing for an equitable transition to the green economy – the role of law), funded by the European Research Council (grant 335592). My thanks to BeneLex PI Elisa Morgera, a great scholar but much more importantly a good friend and fellow traveler from hostels in the Andes to tents in the Kalahari. BeneLex research fellow Elsa Tsioumani, another good friend, led the research in Greece. Elsa, thank you for the hours spent trying to explain Greek and EU law to me; we will get back to Ikaria one day. Other BeneLex research fellows and collaborators have also been kind enough to comment on early drafts and have inspired this work in many other ways – Annalisa Savaresi, Saskia Vermeylen, Kim Bouwer, Margherita Brunori, Wim Peters, Mitchell Lennan, and many others. Particular thanks and credit to Mika Schröder, my fellow in the long, long process of coding and interpreting COP decisions for discourse and frame analyses. Thanks too to Christine Frison and Daniel F. Robinson for helping to improve the structure of the book.

All mistakes and misunderstandings in this work remain, of course, my own.

Friends in Lincoln were a great support during the ups and downs over the years of this project – Ana Jordan, Jan Dobbernack, Adele Langlois, Kaisa Hinkkainen Elliott, and Tina Haux, thank you for the writer's support group! Duncan French and Louis Kotzé have always been kind and encouraging about this work, and patient in explaining points of law to a sociologist with no Latin. Colleagues and friends in Trento have been equally welcoming and even more patient as I tried to get this book written – thanks in particular to Milica Pejovic, Elena Pavan, Anna Casaglia, Andrea Fracasso and Matteo Borzaga, and all my colleagues for their support, both moral and professional. Emma Mitrotta, thank you, for being a friend, a great research assistant, general cheerleader, and proof-reader extraordinaire.

To my family, thank you for being there, as always. Alberto, who is patience itself, and Tom and Daisy, who always believe I can do it.

1 Studying benefit-sharing from the bottom up

This book began as an investigation about the ways that local communities experience benefit-sharing, asking what issues those experiences raised, whether arenas of global environmental governance address those issues, and whether local communities participate in those debates. The guiding principle for the study was to begin from the bottom and move up – in other words, to let local experiences of benefit-sharing provide the parameters for studying global arenas. As is usually the way in empirical research, the study shifted as it unfolded, and opened up reflections which, though they began with local community discussions around benefit-sharing, moved beyond this narrower focus to touch on central themes of fairness, participation, and democratic environmental governance. In this sense, the research presented here contributes to a view of global governance from below, and benefit-sharing emerges as a starting point for an exploration of wider themes at the fore of many different disciplines that seek to grapple with the dilemmas of environmental politics and governance today. The initial focus on benefit-sharing is a window to look at how we might go about including local communities more effectively in seeking and implementing solutions to environmental problems via participation from the local level through to the global.

While the book departs from local-level case studies to identify themes of central importance to local communities discussing issues around benefit-sharing, the findings from those cases raise points that apply to environmental governance in a much broader way beyond this specific policy area. These broad themes then form the basis for a study of the Convention on Biological Diversity (CBD) as one important arena of global environmental governance, and the most developed as far as benefit-sharing is concerned. This study aims to assess where discursive spaces are available for raising the themes that emerged as important for local communities. Moving on from this discussion of discursive space, a more practical analysis of the CBD focuses on paths for local communities to participate in the work of the Convention in different ways. One possible tool for local community participation is discussed in more depth: community protocols. These instruments have garnered attention from scholars and policy practitioners alike (Jonas, Bavikatte and Shrumm, 2010; Jukic and Collings, 2013; Tobin, 2013; Bavikatte, 2014; Bavikatte, Robinson and Oliva, 2015; Delgado, 2016), and have been specifically recognized by the CBD in its Nagoya Protocol on Access and Benefit-sharing (Convention on Biological Diversity, 2010). They form an interesting focus for reflecting on possible direct paths for local community participation. Finally, the book concludes by placing the discussion of local community themes and the spaces they find at the international level into wider discussions underway amongst scholars of environmental politics and civil society. In particular, the findings can be situated within work on participation by civil society and indigenous peoples' organizations in global environmental governance, in work on global civil society, and finally in the discussion of the potential for deliberative democratic models of global governance.

To lay the foundations for all of this, this chapter provides a brief overview of the concept of benefit-sharing and its presence in global environmental law, particularly within the CBD. It also reviews some of the literature on norm diffusion, on local communities, on benefit-sharing, and on global environmental governance from various relevant disciplines including socio-legal studies, environmental politics, political ecology, and work on community-based natural resource management, with a view to underlining the need for a comparative, bottom-up study that focuses on the meanings assigned to issues around benefit-sharing. An overview of the methodology and methods used in the study, which combines qualitative empirical work with quantitative discourse and frame analysis, is also included, along with some brief reflections on the ethical questions linked to local community research. Finally, the chapter presents an overview of the rest of the book.

Understanding benefit-sharing in an interdisciplinary perspective

The research presented in the following chapters was carried out within the framework of a five-year project investigating benefit-sharing both as an evolving concept in different sites of global environmental law and from the perspective of local community experiences.[1] The important point to emphasize for the purposes of this work is that benefit-sharing is not a concept with a fixed and clear meaning. Its definition and implementation (where schemes are in place) vary both across sites of international environmental law (Morgera, 2016; see Savaresi and Bouwer, 2018, on climate change; see Tsioumani, 2018, on agricultural biodiversity), across regional and national regimes, and at the local level. In addition, few functioning benefit-sharing schemes are in place (Robinson, 2015), limiting the scope for research on best practices. Due to this, a gap has emerged between different applications and the principles that benefit-sharing – or fair and equitable benefit-sharing at least – seeks to support. Benefit-sharing attempts to address questions of equity in environmental governance, particularly with regard to developing countries, indigenous peoples, and local communities (Morgera, 2016). In inter-State benefit-sharing (between States), it should involve the rewarding of efforts for environmental protection via, for example, payments, financing, technology transfers, and capacity-building. In intra-State benefit-sharing (within States[2]), benefits such as profit-sharing, the recognition of customary practices, job creation, and other 'goods' should flow to communities that are environmental custodians (ibid.). Yet in practice, many scholars have found this latter type of intra-State benefit-sharing wanting in terms of effective contributions to equity. On the contrary, they have found that it risks reproducing colonial patterns of governance where genetic resources are concerned (e.g. Mulligan, 1999), leads to the imposition of organizational forms on communities that exclude many from benefits (Vermeylen, 2007), or is used as nothing more than a mechanism for the exchange of money between community leaders and external actors in return for consent to carry out environmentally and socially harmful projects (Gilbert and Doyle, 2011).

It is this gap between the theory and practice of intra-State benefit-sharing that drives this research, which departs from studies of local communities. This local-level focus invites further observations about the concept of benefit-sharing, namely that it resembles common practices applied by some local communities for centuries. The rules local communities develop for access to common pool resources – forest products or water sources for example – have been proved to be valid sustainable models since Elinor Ostrom's work (e.g. 1990) decisively challenged Hardin's 'tragedy of the commons' hypothesis (1968), which links open access to inevitable environmental degradation. Commons approaches recall features of what intra-State benefit-sharing seeks to achieve. They are approaches to the sustainable use of natural resources for the benefit of all in the immediate community that may also deliver global goods in terms of environmental conservation and protection. Work on the commons also highlights that in order to work, commons approaches need to exist within wider supportive governance regimes (Ostrom, 1990). Intra-State benefit-sharing goes a step beyond this to envisage more than support and compensation, but positive returns beyond these for environmental stewardship (Morgera, 2016).

The most developed source on fair and equitable benefit-sharing is the CBD, which means it is a good focus to begin to look at how issues linked to benefit-sharing by local communities are reflected at the international level. As explained in more detail in Chapter 4, the CBD has near universal membership, is binding on its parties, and the text of the treaty names fair and equitable benefit-sharing as one of its three objectives alongside the conservation of biological diversity (or biodiversity) and the sustainable use of its components. The Nagoya Protocol to the CBD lays out rules for access and benefit-sharing (ABS) related to genetic resources (Convention on Biological Diversity, 2010). In addition, CBD decisions have progressively come to refer to indigenous peoples and local communities as actors that should be involved in the governance of biodiversity over time. It also recognizes the importance of their traditional knowledge, innovations, and practices in conserving biodiversity in its articles 10(c) and 8(j).

Given all of this, the research presented in this book aims to investigate how issues linked to benefit-sharing are perceived at the local level, before examining where there are spaces for themes important at the local level to be raised at the international level in the CBD. It aims to do this in an interdisciplinary perspective, yet the main focus of the work remains sociological in that it deals primarily with the social meanings attached to issues linked to this international policy in different local contexts. The need for an interdisciplinary approach to benefit-sharing lies in its lack of fixed meaning. Understanding the way that this concept is translated or framed within different social sites, and how the meanings attached to it shift or diffuse, can shed light on its potential for addressing equity concerns. An interdisciplinary approach is also dictated by the observations of legal scholars who place benefit-sharing in the space of global law (see, e.g. Parks and Morgera, 2015). Global law describes a 'pattern of heavily overlapping, mutually connected and openly extended institutions, norms and processes' (Walker, 2014: 11–12) that make up some legal regime. Benefit-sharing, as already noted, has evolved at different levels, sites, and with reference to different natural resources. A range of State and non-state actors are involved in benefit-sharing – international organizations, States, sub-state authorities, civil society actors, and local communities all play a role in implementation (Robinson, 2010). This forms part of recent moves towards decentralized approaches in environmental governance, as the difficulties of reaching agreement via traditional intergovernmental decision-making in multilateral environmental agreements (such as the CBD) have become apparent (Walker, 2014; Stevenson, 2018, for an overview; Mitrotta, 2019). Approaching benefit-sharing through the lens of different disciplines that can shine a light on the roles of different actors at different levels of governance can contribute to a more complete view.

The following subsections review some of the key literature that helps situate this study. First, the literature on norm diffusion from various disciplines helps clarify the focus on socially constructed meanings attached to issues linked to benefit-sharing in local communities. Then, the literature on local communities, benefit-sharing, and global environmental governance is introduced. Though this review is far from exhaustive, it serves to underline the need for a comparative, bottom-up view of local community experiences linked to benefit-sharing.

Norm diffusion – an interdisciplinary overview

Since benefit-sharing involves multiple governance levels, a range of actors, and is attributed varying meanings in different sites, the literature which discusses how legal and political norms diffuse, and how their meanings evolve, appears a good place to start to situate the research. Scholars of law, international relations, political sociology, and other disciplines have paid attention to these questions, investigating the ways that ideas and norms change when appearing in different sites, as well as the mechanisms behind the diffusion of legal, political, and other norms.[3] The present study is limited to investigating the meanings linked to benefit-sharing in different sites, and to identifying any common themes among them. Rather than looking at definitions of benefit-sharing itself, the study looks at the issues linked to benefit-sharing by local communities (see the following section: Studying benefit-sharing from the bottom up). It also focuses on cases where local communities discuss possible intra-State benefit-sharing as opposed to inter-State benefit-sharing. The following brief overview of norm diffusion literature thus pays more attention to work discussing meanings – work on mechanisms of norm diffusion between States is covered in less depth.

In the legal literature, norm diffusion concerns 'when one legal order influences another in some significant way' (Twining, 2004: 5), and is understood by comparative legal scholars using a variety of concepts including 'reception, transplants, spread, expansion, transfer, exports and imports, imposition, circulation, transmigration, transposition and transfrontier mobility of law' (ibid.: 5). This suggests that diffusion is useful for a study that adopts a global perspective to understand benefit-sharing from the bottom up, allowing an analysis of how law moves across borders, different territorial levels, and the roles of a range of actors in this process. However, research on diffusion in the legal literature is limited in general, and empirical studies in particular are lacking. Twining, reviewing the legal scholarship on diffusion, underlines the need to bridge legal scholarship with work across the social sciences to underpin empirical research, since no 'systematic theory' of the diffusion of law exists, and many studies rely on 'naïve' or 'country and western' models where law is transplanted from developed to developing countries (Twining, 2005).

More recent studies buck this trend, and include views of bottom-up legal diffusion and the social situatedness of legal norms. Westbrook, for example, theorizes different scenarios of diffusion: *imperium*, the imposition of law by a sovereign; *fashion*, where law changes according to what is perceived as modern; *system*, where globalization effectively forms a new society with its own legal system; and *tribe*, where law travels from the bottom up, from people to State, and is decoupled from physical space (Westbrook, 2006). Another relevant example is work by Sarfaty (2007), whose empirical, bottom-up research uses an ethnographic approach to explore how the Pimicikamak Cree Nation translated international law into new forms of indigenous law. Her work develops a model of legal mediation where 'a process of negotiation among multiple normative commitments and legal entities' takes place, and 'local actors play an important role in shaping how international norms become internalized within their communities' (ibid.: 444). This research also points to the importance of framing in local redefinitions of law from outside sources. Many socio-legal scholars also touch on questions of norm diffusion in work investigating the ways in which law is applied, implemented, and reframed in local communities (see the next subsection: Local communities and global environmental governance). Framing – a sociological concept – is particularly useful for the present research. It is through this sort of 'meaning work' that legal norms can be re-elaborated in local contexts. In this sense, the lack of conceptual clarity about benefit-sharing could be described as a result of a number of coexisting framings of the same. This contributes to confusion about how to move forward to implement fair and equitable benefit-sharing (see for example McCool, 2012, on the social justice frame of benefit-sharing).

Twining's view about the usefulness of drawing on concepts from the wider social sciences is confirmed by these examples. He singles out the literature on social movements as a particularly useful source, since this is an area where studies of framing are common (Twining, 2005). Hafner-Burton and colleagues (2012) make a similar point, calling for collaborations between political scientists and international lawyers. In sociology, much of the literature on norm diffusion concerns organizations, and discusses diffusion as regards, for example, business models, practices, and institutions. In a review of the field, Djelic argues that sociological studies of diffusion broadly seek to answer the question of social similarity (2008). It assumes that societies are interdependent, and describes diffusion through dense institutional channels in different systems. These channels may be formal and institutional, or made up of individual or collective norm entrepreneurs who act to promote diffusion.

To give a brief historical overview, sociological research moved away from assumptions of efficiency and rationality as the main drivers of diffusion towards the end of the last century. The seminal example of this shift is DiMaggio and Powell's (1983) work on institutional isomorphism, which demonstrates that organizational similarity is not always due to efficiency. Instead, organizational change may occur within a 'logic of consequences' (because of efficiency concerns) or a 'logic of appropriateness' (because of normative concerns), which the authors link to different paths of institutional isomorphism. Strang and Meyer's (1993) work also moves beyond ideas of diffusion as efficiency, and draws attention to the role of theorization, which can drive diffusion by linking disparate actors and providing a motivation for adoption. Many other sociologists investigate the mechanisms rather than the reasons for diffusion. Though mechanisms of diffusion are beyond the remit of the present study, a brief look at this aspect is useful in that it draws attention to the issue of how norms that diffuse 'fit' into new sites, bringing the discussion to framing.

Strang and Soule (1998), to give one example, investigate the different rates and pathways of diffusion in organizations and social movements, underlining that both structural and cultural bases need to be taken into account when considering mechanisms for diffusion. Djelic (2004) reflects on social networks as mechanisms for diffusion, and discusses strong and weak ties within these. In the literature on social movements, the emphasis is on how frames diffuse, but reference is also made to mechanisms. Work in this vein deals with active and passive mechanisms for diffusion, where the former may take place through hierarchy or mimicry (e.g. Snow and Benford, 2009) and the latter can occur through external channels, notably the media (e.g. della Porta and Kriesi, 2009). Through a focus on frames, Benford and Snow highlight that in many cases diffusion is an interactive process, whether actors are directly connected or not, as 'meaning work' is always required for a frame to fit a new context (2000). By shedding light on how norms are diffused in interactive processes, work on mechanisms for diffusion thus moves towards considerations of the *content* of norms and how they are framed or translated in different contexts.

As noted, the literature on social movements is particularly useful here. Studies of social movements focus on collective actors' roles in norm diffusion and underline the diffusion of frames. This concept emphasizes ideas such as translation and the 'fit' of norms in different contexts. Framing was first developed by Erving Goffman, who saw frames as keys used to bring into focus different aspects of situations: a frame or 'a particular definition *is in charge* of a situation' (Gamson, 1985: 686). Different actors frame issues in order to attach characteristics and definitions to people and issues in space and time. They attribute blame, outline alternative paths and means of achieving goals. Frames, in this view, perform the role of interpreting the significance of a person, event, or symbol. This flows from a constructivist view of the social world, where 'meanings do not automatically or naturally attach themselves to the objects, events, or experiences we encounter, but often arise, instead, through interactively based interpretive processes' (Snow, 2004: 384). To frame an object, experience, or event thus requires actors to engage in 'meaning work', where different techniques are employed in the attempt to fit a frame to a context. Techniques discussed in the literature include, for example, articulation (telling a compelling story) and amplification (stressing certain aspects) (Benford and Snow, 2000), which can be summed up with the idea of translating a norm to fit a new setting. Fit, on the other hand, can be understood through the ideas of salience and resonance, which refer to the credibility, compatibility, and coherence of a frame in a new setting (Johnston and Noakes, 2005).

In sum, sociological views of norm diffusion furnish some useful considerations for a bottom-up study of benefit-sharing, particularly in an intra-State dimension. First, they underline that norm diffusion is not linked to superiority or efficiency. Second, studies of framing in social movements point to the possibility of finding common core themes about benefit-sharing across local contexts. As these studies reveal how frames are re-elaborated to fit different contexts, they suggest that some of a norm's content may remain intact to a degree. For legal norms, this idea can be understood through the lens of living law, where the aim of a law is defined by some authority, but the way that this aim is achieved can be adapted to a specific context by the various actors concerned. These actors seek to achieve the aim, but via locally appropriate ways. This view underlines how a study of the framings attached to the 'global norm' of benefit-sharing in different local settings can reveal themes which arise in the diffusion of this norm. These themes can, in turn, feed back into ongoing 'meaning work' around benefit-sharing to refine this norm over time. To explore this last point, a view of how norm diffusion is discussed in the literature on international relations can provide some insights about how local 'meaning work' can affect framings in international arenas.

As a discipline, international relations is concerned with the study of interaction, which means that while diffusion is an implicit theme in much of the work in the field, explicit analyses are more recent (Gilardi, 2012). Scholars of international relations point to a number of diffusion mechanisms at work between States, including coercion, competition, learning, and emulation (ibid.). The last of these, emulation, is discussed in work that investigates why States adopt norms that are not evidently in their self-interest – that is why norms diffuse outside a realist understanding of international relations. Checkel's (2005) work on socialization and a shift from a logic of consequences to one of appropriateness (see above and DiMaggio and Powell, 1983) is one such explanation. He argues that States' strategic calculations, rooted in a logic of consequences, may over time be internalized, which means that a norm subsequently spreads in a logic of appropriateness (where a State adopts a norm because that is what the State in question 'has always done'). Another scenario is when States adopt a norm because to do so will simplify a task. This can occur regardless of whether the State has internalized the norm in question – it is a reputational move rather than a move based in real beliefs. Even so, this is described as a first step towards switching from a logic of consequences to one of appropriateness, as internalization of the norm may follow adoption. Finally, in what he terms a scenario of normative suasion (drawing on Habermas' idea of communicative action, see Chapter 6: Local communities, participation, and democratic environmental governance; and Habermas, 1984, 1987), State agents may 'actively and reflectively internalize new understandings of appropriateness' – that is they are *convinced* that the new norm is simply right (Checkel, 2005: 812).

A complement to this idea of norm diffusion among States in international relations is Finnemore and Sikkink's (1998) model of the *stages* of diffusion. In an initial stage of 'norm emergence', norm entrepreneurs propose a new norm. If and when enough actors take up a new norm, a 'tipping point' is reached and the 'norm cascade stage' begins, where non-compliance with a norm is punished. Finally, the 'internalization' stage is reached when a norm is no longer questioned (the authors use the examples of slavery and universal suffrage to illustrate this point).[4] This is not necessarily the destiny of all norms, however – norm diffusion is not inevitable and may well be a lengthy process (ibid.). Implicit in these views of diffusion is the idea that norm entrepreneurs are powerful, and that norms constitute social hierarchies. Yet norms sometimes originate in less powerful States, or from the bottom up. Towns (2012) discusses the diffusion of norms originating from States in Latin America, often seen as norm recipients rather than entrepreneurs. These States, perceived as 'lower down' in an international hierarchy, may seek to introduce new norms in a bid to improve their standing (ibid.: 183). Here, the framing of the norm is identified as a crucial component, as States may draw on internalized norms to promote new ones.

These views of norm diffusion in international relations echo many of the points made in sociological studies, but convey an idea of how norms framed within States may then be introduced in international relations, and how this may be a desirable thing for States perceived as lower in international rankings. This is a useful idea for the present study, which seeks to shed light on the spaces for themes framed by local communities as important to benefit-sharing at the international level. Though a study of how and whether such local norms spread to the international level is beyond the scope of the research presented here, this literature provides the basis for a reflection on potential paths for local to international norm diffusion (see Chapter 5). More generally, the many similarities and overlaps between these bodies of literature suggest that the research in this book contributes to a number of disciplines. The aim is to answer calls for a more unified approach to address single ideas through the lens of multiple disciplines. The following subsection integrates this multidisciplinary background to the research with a review of literature from various disciplines on local communities and global environmental governance.

Local communities and global environmental governance – a non-exhaustive overview

The focus of this book is the themes identified by local communities around the idea of benefit-sharing, and where these issues find spaces at the international level. The previous subsection reviewed literature about the ways that norms move in societies and between States. This subsection turns to literature that looks at discourses in global environmental governance, and the governance of biological diversity especially, as well as how these discourses affect local communities. A main focus on biological diversity is motivated by the need to delimit the review, as well as the fact that this is the area of global environmental governance with the most developed benefit-sharing regime. In addition, the analyses of the international level in this book focus on the CBD (see Chapters 4 and 5). The review once again spans different disciplines, including international environmental law, environmental politics, socio-legal studies, and political sociology. All of the scholarship reviewed is united by a focus that chimes with political ecology – viewed as a multidisciplinary collection of studies that aim to understand the ways in which people and societies interact with natural resources, which are seen as embedded in political, power-laden contexts (Robbins, 2012). In the following, literature on global environmental governance relevant to exploring how discourse shapes the governance of biological diversity and benefit-sharing, and consequences at local levels, is introduced. The subsection ends with a brief overview of literature on community-based natural resource management (CBNRM), which argues the importance of real power transfers to the local level for effective environmental governance, despite empirical findings that are often critical of existing schemes.

Much of the work reviewed here shares an underlying view that norms matter, and shape political spaces such as that of global environmental governance. This occurs via discourses, which have been argued to underpin norms. The term discourse is assigned various meanings in the social sciences. Here, and in the literature reviewed, discourse is understood in a Foucauldian perspective. Moses and Knutsen (2012) provide a useful overview, noting that the Foucauldian definition of discourse grew from this author's idea of mentalities, which are collective understandings held by a group in society that dictate criteria for judging different types of behavior. As mentalities are repeated over time, they become more deeply rooted and thus form discourses, which ultimately delimit and shape political action by defining what is considered acceptable, sensible, and normal in a society (ibid.: 217). The discourses seen as shaping global environmental governance, including the governance of biological diversity, are often rooted in Marxist critiques of international political economy, which view liberal market capitalism as a hegemonic discourse that emphasizes the importance of private access and control over resources (Ribot and Peluso, 2003; Svarstad, 2004). This emphasis on discourse – if we accept this as underpinning norms – bridges the review of the literature here with the point raised in the previous subsection about hierarchies of norms, and norm entrepreneurs.

A first area of literature relevant for this research investigates how discourses shape arenas of global environmental governance themselves. Reimerson (2013) develops an argument about how colonial discourse continues to shape global environmental governance and analyzes the extent to which the wording of the CBD creates space or, conversely, limits the agency of local communities and indigenous peoples. First, Reimerson (ibid.) notes that there has been increased recognition of the role played by indigenous peoples and local communities, and indeed traditional knowledge, in the protection of biodiversity and the environment. Despite this, widespread threats to these groups' lands persist, while the text of the CBD perpetuates some colonial discourses that limit the scope of indigenous peoples and local communities to draw on it to protect themselves from these threats. Specifically, she notes that indigenous peoples and local communities continue to be 'Othered' in the text of the CBD – a practice echoing the actions of colonial forces. Colonialism not only 'Othered' nature, defining it as a wilderness to be tamed, but also 'Othered' its residents, equating them with 'wild' nature and ignoring their roles in shaping lands and waters. These processes were often justified through scholarly work such as the theory of the tragedy of the commons (Hardin, 1968; see Said, 1978).[5] This attitude, she finds, is perpetuated to some extent in the CBD's recognition of indigenous peoples and local communities only insofar as their traditional lifestyles and knowledge contribute to protect biodiversity, limiting their space for action. However, the 'nature–culture dichotomy' that underpins colonial discourse – that is the distinction made between 'wild' nature and the role of man – is less apparent in the CBD. This leaves some space for the agency of indigenous peoples and local communities, and Article 8(j) in particular is recognized as a basis for action, albeit conditioned by the affirmation of State sovereignty (ibid.).

The consequences of the definition of what is 'natural' and how this is understood as separate from 'culture' is investigated with regard to both the CBD and the United Nations Framework Convention on Climate Change (UNFCCC) by Uggla (2010). Beginning by noting that nature has long been 'Othered' in discourses rooted in Judeo-Christian traditions, Uggla identifies the motivations and meanings behind nature in these two treaties. The CBD is seen as unique in its recognition of the intrinsic value of biodiversity and the various meanings of nature,[6] while the UNFCCC is understood to address climate change only insofar as it affects human life. Uggla (ibid.) concedes that intrinsic value may be an unwieldy idea for actual policy-making, but concludes that '[b]y treating humans and nature as inter-related, discussion of environmental protection and social justice could focus on how to reduce human vulnerability and accomplish sustainable living conditions, instead of being caught in futile negotiations around how to define and distinguish natural and human impacts' (87). In this view, the underlying discourse shapes decisions about how to tackle environmental problems.

Another fundamental definition – this time of biodiversity itself – is discussed by Escobar (1998). Biodiversity, he argues, is itself a construction that 'articulates a new relation between nature and society' (55). By pointing to a new relation, biodiversity inspired the definition of new discourses, none of which were hegemonic initially. Discourses were produced and disseminated by different actors as the term biodiversity emerged. A first group of actors comprised those working on conservation biology and fed into the Rio process in the early 1990s. They advanced different discourses, including resource management, which sees threats to biodiversity as linked mainly to habitat reduction, and thus as resolvable through management mechanisms. A secondary discourse in this group is identified around sovereignty over genetic resources and transfer processes, and is attributed to third world States (ibid.). A further two discourses are attributed to southern non-governmental organizations (NGOs) and social movements respectively. Biodemocracy, attributed to the former, blames threats to biodiversity on the root cause of habitat loss, identified as consumption habits in the global North. The solution is local control over natural resources, and redefinitions of productivity and efficiency. Finally, social movements 'explicitly construct a political strategy for the defense of territory, culture, and identity linked to particular places and territories' and view the construction of the term biodiversity hegemonic, but also perceive a space for construction (ibid.: 60). This view of discursive contestation fits the approach adopted by Dryzek (2005), who maps discourses of environmental politics at a more general level and points out that not all of these are necessarily incompatible. Indeed, Dryzek sees radical green discourse – whose proponents include social movements – as providing much needed provocation and input for more management-oriented solutions, an observation that could be applied to Escobar's view of biodiversity (ibid.). Dryzek's work sets an important frame for a central research question in environmental politics – to what extent are the different understandings of nature and social organization held by different groups in society incompatible, and to what extent can they be combined to address environmental problems?

Other work relevant for the research presented here discusses the agency of indigenous peoples and local communities[7] when exerted directly in arenas of global environmental governance. The intergovernmental Conferences of the Parties (COPs) of various treaty bodies, where States that have signed a treaty meet periodically to discuss implementation, are a major focus here. Marion Suiseeya (2014), drawing on the results of a collective ethnography carried out by a team of researchers, discusses the negotiation of the Nagoya Protocol in this view. Focusing on whether the negotiations tackled questions around meanings of justice, and allowed indigenous peoples and local communities to 'introduce new ideas, values, and norms that, over time, may shift the value orientation of global environmental governance' (ibid.: 103), she finds that negotiations were instead dominated by debates over instruments. Acknowledging that the Nagoya Protocol nevertheless signified a step forward for indigenous peoples and local communities, she finds little discursive space for negotiating fundamental meanings. The key to understanding this limitation is the presence of a 'meta norm' that places 'justice practices' in the instruments of prior informed consent and mutually agreed terms (ibid.: 118). Brand and Vadrot (2013) argue in a similar vein that 'epistemic selectivities' in the CBD impose limitations about which arguments are admissible, excluding, for example, traditional knowledge outside the requirements of the Nagoya Protocol. The acceptance of these instruments limits debates to questions of good and bad practices within their definitions, and pushes more fundamental questions aside. In the UNFCCC's Paris Climate Summit, however, Marion Suiseeya and Zanotti find evidence of more effective agency exercised by indigenous peoples' organizations by broadening their definitions of power and influence (2019). However, earlier meetings of the UNFCCC have been judged to provide less space for the agency of indigenous peoples and local communities (Eastwood, 2019).

Bavikatte's (2014) work also discusses the agency of indigenous peoples and local communities in the negotiations of the Nagoya Protocol. His wider study is an investigation of claims for biocultural rights,[8] which he locates in the 'confluence of movements for post-development, the commons, and the rights of indigenous peoples' in a political ecological approach (ibid.: 16). This movement, he argues, is faced with laws that reify hegemonic, capitalist market-based discourses, and privilege the Lockean conception of private property as a right of exclusive access and use. He sees the achievements of the representatives of indigenous peoples and local communities in shaping the text of the Nagoya Protocol as a notable challenge to this. In particular, he argues that the recognition of community protocols (where local communities document information on their practices, institutions, customary law, and aims, among other things) is a means for the recognition of different property regimes based in non-market worldviews.

Johnson (2014) takes a novel approach to the question of indigenous agency in arenas of global environmental governance, and examines a moment when Inuit discourses linked to food sharing affected decision-making at the international level. Using the lens of affect, which broadly speaking admits emotion to the realm of rationality to overcome 'the splitting of mind and body that is characteristic of western thought traditions' (ibid.: 166), Johnson analyzes how discourses of food sharing within an ethic of care were used in the interventions of the Chair of the Inuit Circumpolar Council at international meetings to negotiate a Global Convention on Persistent Organic Pollutants. The ultimate effects of such interventions are, however, judged to be negligible, particularly given the findings of literature focusing on how hegemonic discourses shape local communities.

In that vein, Mulligan (1999) provides an early consideration of the implications of the CBD's benefit-sharing commitments for the local level, drawing on case studies submitted to the fourth Conference of the Parties (COP) (before the adoption of the Nagoya Protocol). Mulligan departs from the observation that the model of access and benefit-sharing contained in the Nagoya Protocol recalls colonial relationships, a point also argued by Reimerson (2013). The idea of the nature/culture divide also emerges from his discussion of the potential incompatibility between traditional knowledge and the market model in this instrument. Commercializing traditional knowledge, he notes, is likely to have detrimental effects on the social structures that produced such knowledge in the first place (Mulligan, 1999: 47; Vermeylen, 2017). Once knowledge is shared, other problems may arise, particularly if those that held the knowledge originally are dispersed. On this point, Vermeylen documents the array of problems encountered after the conclusion of a benefit-sharing agreement on Hoodia with the San people of Southern Africa (2007). The creation of institutions that failed to reflect the breadth of this community caused various issues, not least the ineffective communication of information about the agreement throughout the group. Robinson (2015) details a wide range of local community cases of access and benefit-sharing linked to the Nagoya Protocol. These cases underline the particular circumstances that each case of traditional knowledge presents, and the different ways that benefit-sharing may unfold. Cases where local communities are more involved in the development of benefit-sharing agreements emerge as those achieving greater success in terms of local communities sharing equitably in benefits, though all of the cases raise points for further practical consideration by policymakers (ibid.).

Adger and colleagues (2001) also underline local-level specificities. They investigate the assumption that environmental problems can best be solved by globally negotiated and implemented approaches. This view, they argue, unites a number of discourses that characterize the Rio treaties.[9] Their work focuses on local-level case studies and challenges this view of the efficiency of global approaches given the complexities of local realities. Regarding biological diversity, the authors identify a 'win–win' discourse as underpinning the CBD, and benefit-sharing from access to genetic resources in particular. In this view, benefit-sharing with the providers of genetic resources is argued to allow all parties involved to profit. Nevertheless, this discourse perpetuates a colonial view of global North–South relations, as bioprospecting (the export of plant material from the global South to the global North) originated in this era. Exports of genetic material contributed to fuel further colonial and economic expansion, leading to the redefinition of bioprospecting as biopiracy (ibid.; see also Shiva, 1997; Parry, 2004; Robinson, 2010; Oguamanam, 2012). The CBD's objective of benefit-sharing seeks to create fairness within these exchanges, but nevertheless reproduces the same general model, which leads to the imposition of this approach in too uniform a manner. Indeed, the authors evidence this through a range of local case studies which reveal complex and nuanced situations in bioprospecting relationships, not least concerning the understandings of benefits and their distribution. These cases highlight the shortcomings of the discourses associated with the CBD, and its limitations in addressing local particularities from the global level (Adger *et al.*, 2001).

Other global environmental policies also receive attention from scholars interested in local effects. Wilkinson (2014) explores payments for ecosystem services through an eco-feminist framework, focusing in particular on critiques of western rationalism, the systemic consequences of ideology, and the materialist implications of both. Western ideology has its 'basis in logic structures that continue to reinforce the domination, marginalization and a dualist separation between the "valued" and "devalued"' (ibid.: 172), recalling points about colonialism underlined earlier, but extending the argument to note that masculine ways of using the body and the environment for economic gain are generally privileged in western discourses. These discourses play out in the view of value linked to market economies, for example the dominance of private property regimes based on the right to exclude (Bavikatte, 2014). The material effects of this narrow definition of value can be read in the lives of women whose work tends to be devalued in ecosystem services payment schemes such as REDD+[10] (Castree, 2008; Wilkinson, 2014). In line with other work reviewed here, Wilkinson finds schemes underpinned by discourses that threaten the systems that produced the 'good' they seek to protect – such as sustainable forest use or the production of useful traditional knowledge – problematic (see also Natarajan and Khoday, 2014).

A final relevant area of literature, which again spans different disciplines, focuses on benefit-sharing of different forms via local-level case studies, underlining that benefit-sharing can take place in a range of approaches. Wynberg and Hauck underline this point in a volume gathering case studies of benefit-sharing in South African and Mozambican coastal communities (2014). The cases include benefit-sharing through CBNRM, co-management, revenue sharing and corporate social responsibility schemes, and concern a range of different natural resources. Despite this variety, the cases suggest some common themes of importance for evaluating benefit-sharing that echo the findings of this book (see Chapter 3 in particular). The authors draw attention to the need to consider the ranges of different actors involved in benefit-sharing, the power held by each actor and over what, as well as the political, economic, and legal contexts they act within. They also argue the importance of the processes and institutions used for developing benefit-sharing, including, crucially, the definition of benefits (see also Tysiachniouk and Olimpieva, 2019). The emphasis for guaranteeing good outcomes in benefit-sharing, in this view, comes *before* any actual agreement is in place and suggests the importance of factors that enable benefit-sharing (Wynberg and Hauck, 2014: 155), as well as the question of addressing power balances between different actors (see also Swemmer *et al.*, 2015, for a practitioner view).

Questions of power and institutions also come to the fore in the literature on CBNRM. This literature includes a great many community-level case studies, which are particularly informative for a bottom-up approach to studying of benefit-sharing. Work in this vein also finds that the discourses argued to underpin global environmental governance affect local communities, for example by underlining the damage done by colonial models of 'fortress' natural resource management. This approach justified the displacement of entire communities and peoples in a logic that equated them with 'wild' nature, denying their role in shaping and conserving landscapes (see Uggla, 2010; Reimerson, 2013). Child and Barnes provide an overview of the concept of CBNRM, defining it as 'a rigorous process of institutional reform that combines the devolution and delineation of property rights with collective action in rural communities to improve the value and sustainability of wild resources' (2010: 283). In addition, CBNRM (and similar approaches) unite the objectives of environmental conservation, rural economic development, and rural institutional and organizational development (Murphree, 2004). CBNRM thus seeks to reconcile goals held important by rather different groups of actors: scholars, NGOs, and funders, who value conservation; and States and local communities, that value rural economic and institutional development to compensate the costs of conservation costs and generate income (ibid.). However, it is the democratic potential of the concept, which foresees the devolution of power to local levels, which makes it particularly attractive to many communities and scholars (ibid.). Nevertheless, most case study work on CBNRM underlines its failures in real-life implementation.

Case studies of CBNRM emphasize the importance of institutions and actors, yet scholars also point out that apolitical readings of CBNRM are problematic for both implementation and assessment. Nelson (2010b), introducing a volume of case studies on CBNRM in Africa, argues that the 'political logic of state control over lands and resources shapes natural resource governance patterns across Africa. Agricultural policy in Africa's largely agrarian nations has evolved according to political interests bent towards controlling producers' access to markets and inputs in order to extract rents' (ibid.: 14). In this context, he goes on to note, local communities and NGOs that support CBNRM are constrained in terms of the extent to which they can secure real transfers of power to local levels to 'deliver more equitable and sustainable resource governance arrangements' (ibid.: 21). The case studies contained in his volume discuss various problems in real-life CBNRM scenarios. For example, they review cases where CBNRM has been subject to local elite capture, which prevents benefits from being shared in local communities in fair ways; instances where NGOs have attempted to apply CBNRM approaches without deep knowledge of local and national political circumstances (Rihoy and Maguranyanga, 2010); and examples where CBNRM schemes come up against discursive power imbalances including clashes between local approaches to conservation and market economy discourses, or the definition of knowledge as western science (Madzwamuse, 2010; Ngoitiko *et al.*, 2010).

However, the potential of CBNRM is also underlined in his volume. The cases of CAMPFIRE in Zimbabwe and community conservancies in Namibia in particular are identified as schemes that have produced positive examples (Child and Barnes, 2010; Jones, 2010). The success of these schemes (albeit variable) is attributed to a range of factors. For Child and Barnes, this is linked to applications in small and relatively homogenous communities (2010). This allows deliberative, participatory forms of democracy to be used as the initial basis for CBNRM decision-making, prior to the introduction of representative democratic elements: such a participatory starting point is deemed crucial to eventual success (ibid.). This observation refers in particular to successful CBNRM examples from Zimbabwe's CAMPFIRE program (ibid.; see also Jones, 2006). Another factor in successful CBNRM schemes concerns institutional backdrops consisting of clear legislation and strong actor networks. This is exemplified in the CBNRM approach applied in Namibia, where legislation was championed by a proactive Ministry of Wildlife, Conservation and Tourism, and local Namibian NGOs worked closely with International NGOs (Jones, 2010).

To reconcile these views, Murphree's observations about the different aims and methods of actors involved in CBNRM help disentangle some of the reasons for the varying success of CBNRM (2004). First, Murphree underlines that funders and NGOs tend to work on the basis of projects, which necessarily set a beginning, end, and a 'goal' to CBNRM schemes. These criteria may be impossible to reconcile with true community-led approaches, which are longer term in outlook and less goal-oriented (ibid.: 204; see also Choudry and Kapoor, 2013). Tensions also subsist between the more economic aims of States and those of local community actors. Here, Murphree uses the example of CAMPFIRE in Zimbabwe to illustrate 'the unintended effect of inhibiting devolution' (Murphree, 2004: 206). In Zimbabwe, CBNRM schemes were developed in the image of a similar approach applied to foster conservation in the private sector. This scheme saw dramatic results in terms of both conservation and economic gains (ibid.). When developing an approach that would allow a similar outcome for communal lands, however, powers were transferred to rural district councils rather than the 'communal entities [...] analogous to the farms and ranches' handed power in the original scheme (ibid.: 206). As revenues were on the increase, neither the rural district councils nor national State actors had any incentive to devolve power to the communities that were effectively managing the land and natural resources. This led to local elite capture in some cases (ibid.). Both Murphree (ibid.) and Nelson (2010a) argue that CBNRM can work, but underline that this requires holistic approaches that take account of political contexts. In addition, CBNRM cannot be 'parachuted' in without regard for local specificities, and must be integrated with other national approaches. Perhaps most importantly, CBNRM schemes must effectively transfer power and provide capacity-building to the local actors that actually manage natural resources. This transfer of power to local levels is also crucial for achieving the participatory democratic element contained in CBNRM. Similar arguments are also made on the basis of case studies of participatory conservation outside the African context (see Bixler *et al.*, 2015).

This overview of literature relevant to studying benefit-sharing from the bottom up is by no means exhaustive. It is limited to areas of scholarship that chime with an interest in a local to international perspective, and in the ways that norms and discourses shape local and international arenas and experiences. The overview began with a focus on norm diffusion in three different disciplines – law, sociology, and international relations. Legal scholars of norm diffusion tend to draw on concepts from the broader social sciences to deal with complexity in their field. Sociological views were shown to be particularly useful with regard to how norms change in different contexts, and the literature on framing provided reflections about how this translation takes place. The sociological literature also underlines the complexity of diffusion, which takes place through different individual and collective actors, on the basis of different logics, and in different directions including the 'bottom' up. The international relations literature draws further attention to logics of diffusion, and explains how local collective action may trigger processes of norm diffusion that can provoke the adoption of norms at the international level for moral reasons (within a logic of appropriateness).

The second area of literature investigated looks at how discourses shape arenas of global environmental governance. A number of relevant points were explored. First, a clear view about the dominant discourses that shape global environmental governance emerges from the literature. These dominant discourses exercise power by shaping worldviews and excluding other minority discourses, and have material implications through their crystallization in policies and law. The central dominant discourse identified in the literature is a capitalist or market-based worldview, which is understood to shape policies in ways that privilege economic gain above other goals. Market-based relations between States, and among communities within them, are argued to damage many local approaches to social organization that do not privilege economic gain. This central capitalist discourse builds in turn on an understanding of a clear division between 'nature' or 'the environment' and 'society' or 'culture'. This means that 'nature' is considered a resource at the service humankind, which forms another threat to systems that regard humankind and nature within a single community rather than dualist perspective. Ironically, holistic views tend to lead to social systems that are better for environmental protection, and create 'goods' that global environmental governance seeks to protect (Vermeylen, 2017). A further discourse that characterizes global environmental governance according to the critical literature is colonialism, linked to capitalist market expansion. Colonial relations between countries in the global North, which exploited the raw materials of the global South, are perpetuated in many ways in the present.

To explore the power relationships at work in discursive contestation, scholars focus on how local communities, indigenous peoples, and other collective actors use their political agency in arenas of global environmental governance to challenge hegemonic discourses. Here, different views of the agency exercised by indigenous peoples and local communities emerge, with some authors more hopeful about their influence than others. On a more abstract level, the literature notes the speed with which our understandings of environmental problems change, and the existence of spaces of discursive contestation to define these (Escobar, 1998). Dryzek's (2005) work provides a broader frame for this view of discursive contestation and underlines the importance of seeking to understand whether and where contestation takes place, and whether certain discourses can coexist rather than clash. Other scholars conduct empirical research to look at how indigenous peoples and local communities challenge discourses in arenas of global environmental governance. Their conclusions about the possibility for these groups to challenge discourses vary according to the area considered (the review focused in particular on the CBD and, to a lesser extent, the UNFCCC), the moment, and the particular subject area. Studies focusing on COPs found less space for agency compared to studies with a thematic focus on a particular area where recognition can be built on by local communities and indigenous peoples (e.g. Reimerson, 2013; Bavikatte, 2014; Johnson, 2014). Overall, this suggests that while there is a generally accepted view of the discursive 'map' of global environmental governance, there is less agreement about when and where discourse can be contested effectively, and where minority discourses can find spaces within the dominant view.

Scholarship focusing on the local level and how global environmental governance is experienced in local realities is also pertinent to the research presented here. Relatively few studies in this area discuss benefit-sharing, but those that do highlight that taking local contexts into account – political, physical, legal, etc. – is crucial. While Adger and colleagues (2001) underline local specificities as a reason to question a global approach to environmental governance, Wynberg and Hauck (2014) find that some common elements emerge with reference to benefit-sharing in their comparison of a range of local experiences (see also Robinson, 2015). This book builds on this latter view in its approach. A last area of literature on CBNRM provides a good source of local community-focused scholarship that includes questions linked to benefit-sharing. Similar arguments about the need for any policy developed outside a community to be applied in ways that account for local specificity exists in this literature. In addition, work on CBNRM makes an explicit link between power, collective action for the environment in local communities, participation in decision-making by local communities, and democracy. Without democratic approaches, environmental policies are unlikely to achieve their aims according to CBNRM literature (e.g. Murphree, 2004; Child and Barnes, 2010). In light of all this, the following section outlines the methodological approach adopted in this book to explore benefit-sharing from the bottom up in a manner that aims to build on these bodies of work.

Studying benefit-sharing from the bottom up: methodology and methods

Building on the scholarship outlined in the previous section, this book aims to look at local community discussions of issues related to benefit-sharing in a comparative perspective. The decision to take a bottom-up approach is motivated by a number of factors. First, benefit-sharing is not a precisely defined policy in global environmental governance, but it is appearing in an increasing number of arenas (Morgera, 2016). Second, much of the literature on local agency in global environmental governance focuses on how dominant discourses can be challenged by indigenous peoples and local communities at intergovernmental negotiations (COPs), and often lacks bottom-up analysis of what indigenous peoples and local communities are aiming to achieve. Third, scholarship focusing on local community case studies highlights the importance of paying attention to local contexts and local-level democracy and participation for understanding successful applications of policy.

The purpose of the methodology briefly described here is to guide a study of how local communities perceive benefit-sharing which reveals the themes they identify as important for its eventual achievement. In the terms of the discussion of norm diffusion, how do local communities frame benefit-sharing? The themes that emerge can then be used to look more closely at the global environmental governance arena of the CBD. The aim of this part of the research is to see if and where there are discursive spaces available for themes identified as important at the local level. The analysis looks at how discursive spaces evolve over time, rather than single intergovernmental negotiations. A second step in the analysis takes a more practical view, exploring how local communities might be able to participate in global environmental governance in more everyday ways. This is not to argue against the importance of studying the agency of these groups during COP negotiations – rather the aim is to shed some light on other, additional, modes of participation that can complement action taken at COP negotiations, but tend to be understudied (Fung and Wright, 2003). Indeed, the paths of participation that do exist for local groups in global environmental governance are partly the result of activism during negotiations. Community protocols, which form one focus in the discussion of participation beyond COPs, are a good example of this (Bavikatte, Robinson and Oliva, 2015). The following subsections outline the methodology, and touch on the methods employed as well as ethical questions with respect to each of these aims. The first is dedicated to the local community case study approach and the comparison of the cases. The second provides an overview of the methods used in analyses of discursive spaces for local themes and participation in global environmental governance.

Comparative local case studies

The basis for discussions about the global level of environmental governance in this book is a set of five comparative local-level case studies.[11] Some words about the underlying assumptions of the methodology are useful to situate this local case study research, and why it is used to feed into reflections on the global level. As already highlighted, a core interest in this book is about the meaning of benefit-sharing – how it is framed by local communities, and how those framings fit into the overall discourses that characterize global environmental governance. The departure point for the research is thus constructivist in the sense that meaning is viewed as socially constructed in interactions between various actors. Meaning is not fixed; rather it shifts over time and context. A case study methodology is a clear candidate for gathering the kind of data that inform about how different actors discuss and frame benefit-sharing in social interactions, and is necessary given the relatively limited amount of information about local-level views available. Case studies are conducive to detailed research about a particular site, and allow attention to be paid to both social interactions and the contexts in which these interactions are situated. The case study approach adopted here is also exploratory, in the sense that no hypotheses about how benefit-sharing is framed in local communities were developed for testing during empirical research. Rather, the idea was to let local communities speak for themselves without interventions from researchers to investigate predetermined hypotheses, and for their views to form the basis for further investigations.

The overall goal of the local-level research was to see whether any core themes characterized local discussions around benefit-sharing. This goal dictates a comparative approach, which raises a dilemma all social science research endeavors face – the trade-off between depth and breadth. The more detailed the case study, the more challenging it becomes to study larger numbers of cases, and to compare them. Yet small numbers of detailed cases cannot be generalized to other contexts or situations. The research in this book requires relatively in-depth case study work. To attempt to address the generalizability issue, five case studies were selected (a reasonable number for a five-year research project) following a most different systems design, where maximum variation is sought for independent variables. The five case studies were chosen on the basis of variation in terms of their location, governance contexts, the natural resources at stake, types of communities, and stages of community discussions linked to benefit-sharing.

To manage to carry out five relatively detailed case studies, partnerships were formed with NGOs working in long-term relationships with local communities discussing issues around benefit-sharing. To identify local communities engaged in discussions around benefit-sharing, community protocols were used as a selection criterion. All of the local communities selected had considered, were drafting, or had already published, community protocols. Community protocols are discussed in another part of this research in their own right as a possible path to international participation (see Chapter 5: The view from below – community protocols and international participation). For the purposes of case selection, however, discussions of community protocols were simply considered as a good indication that a community was discussing issues related to benefit-sharing. Using community protocols as a proxy meant that the chosen case studies included local communities that had not dealt with benefit-sharing as a policy area as such, and their views of the policy were mostly unaffected by the framings of global environmental governance.

The local community case studies chosen are located in Greece, South Africa, Namibia, Argentina, and Malaysia. The Greek case study focuses on a community of traditional goat pastoralists on the island of Ikaria. The pastoralists are facing a range of issues linked to benefit-sharing and access to land against a complex backdrop where European Union (EU) rules and climate change have contributed to problems of overgrazing, amongst others. The South African case concerns a group of traditional health practitioners from a range of ethnic and linguistic backgrounds. This group formed the Kukula Traditional Health Practitioners Association and drafted a community protocol in 2009 (Traditional Health Practitioners of Bushbuckridge, 2010) to highlight their work in communities and threats to biodiversity in the area, adjacent to Kruger National Park. The practitioners, who initially came together to discuss ABS issues (Sibuye *et al.*, 2012), are currently discussing a range of issues including access to medicinal plants and the preservation of cultural heritage, sustainable management and stewardship of land, and participation in protected area management. The Association thus decided to renew its community protocol to reflect these changes. The Namibian case also concerns cultural reproduction in terms of conserving knowledge for future generations, this time in connection with wildlife management and traditional knowledge among Khwe residents of Bwabwata National Park. Historically, this group hunted and gathered food in what is now a core area of the park. In recent times, the communities feel that their traditional knowledge may disappear. They have drafted a community protocol to support their work on cultural reproduction and to underline their role in wildlife management, tied to work taking place through their involvement in the Kyaramacan Association which aids in park management and distributes hunting concessions (see Koot, Beek and Diemer, 2016).

The Argentina case concerns communities resident around the Salinas Grandes and Olaroz salt flats in the Andes. Mining companies have begun extracting lithium, or have engaged in exploratory work, in the area. Some community members from the Olaroz area have raised concerns about consultation and consent process, while a group of 33 communities from the Salinas Grandes area have published a community protocol which also focuses on consultation and consent (Comunidades Indígenas de las Salinas Grandes y Laguna de Guayatayoc de Jujuy y Salta, 2015). The Malaysian case focuses on the Kelabit community of Bario, Sarawak, and Bario rice, a unique variety. An aging population, increased social mobility, and labor shortages have left cultivation levels critical and Bario rice is under threat. An association of Kelabit community members, mostly living away from Bario, secured subsidies from the Sarawak State government for a tri-partite management deal, involving the association, the State government, and a commercial farming company, to grow the rice. The scheme has raised questions around traditional and industrial farming techniques, and a possible future community protocol is under discussion.

As mentioned, the research for the five case studies was carried out in partnerships with NGOs. This means that the research findings are to some extent presented through the lens of the relationships between these NGOs and the local communities, and this should be borne in mind when considering the results. As discussed in Chapters 5 and 6 in particular, NGOs have been argued to shape local communities in various ways through their work with them, particularly if they are unaware of or insensitive to local specificities and political contexts (a point also underlined in the CBNRM literature outlined in the previous section). The NGO partners involved in the research presented here are instead involved in long-term relationships with the local communities studied, and work towards broad aims of empowerment and support for local communities rather than the implementation of specific projects. In South Africa, Namibia, and Malaysia the NGO partner was Natural Justice (an NGO with a great deal of experience in working with communities to develop community protocols), in Greece the NGO partner was the Documentation, Research and Action Center of Ikaria, and in Argentina the Foundation for the Environment and Natural Resources (Fundación Ambiente y Recursos Naturales).[12]

The choice of methods used to carry out the case studies also sought to respond to the aim of gaining an 'organic' view of how issues linked to benefit-sharing were discussed, to the practical issues raised by carrying out five different case studies while working with NGOs, and a range of ethical questions related to research in marginalized communities, consent, and benefit-sharing. At the research design stage, a menu of methods deemed the least intrusive and workable was selected (Morgera and Parks, 2014). Inspired by work on participatory action research, the methods used varied for the different cases after consultations with the partner NGOs. Where possible, participant observation of community meetings was preferred, after consent to observe had been asked of the local community members involved. This method was used in South Africa and Namibia, where it was possible to arrange research visits in conjunction with planned community meetings. In Malaysia, Greece, and Argentina unstructured group and individual interviews were used, as meetings were either not planned to take place, or community members did not wish their meetings to be observed. The notes from participant observations and unstructured interviews were supplemented wherever possible with conversations with NGO partners and semi-structured interviews with representatives of relevant authorities, as well as published sources.

Ethical questions about consent and benefit-sharing also arose as the research was carried out. A consistent common theme for the local communities was indeed about consent (see Chapter 3: Procedural guarantees), and this also affected the way the research was carried out. The uniform approach to seeking consent agreed during the research design stage was supplemented in varying ways for each case after consultations with NGO partners. In addition, efforts were made to share relevant benefits from the research process as the project unfolded. These efforts were also inspired by participatory action approaches and their attention to the co-production of knowledge based on community needs (Morgera and Parks, 2014; Parks and Morgera, 2019). As mentioned in the introduction to this chapter, the case study research took place within the framework of an interdisciplinary project. The project included research benefits aimed at local communities in different ways, including training modules based on the legal findings, as well as direct presentations of interim findings to interested communities. This was made possible by visiting each community (with the exception of Namibia) twice, but more importantly by building reflexivity into the project. Specifically, the case study research inspired work on the interface between prior impact assessments, consent, and benefit-sharing (Morgera, 2018; Marchegiani, Morgera and Parks, 2019), while the international-level research was also driven by the findings that emerged from the local case studies.

In order to guide the case comparison, Ribot and Peluso's theory of access (2003) was useful to structure attention to framing and context. In this approach, access is defined as 'the ability to benefit from things' (ibid.: 153) as distinct from ownership through a property regime. This drives 'grounded analyses of who actually benefits from things and through what processes they are able to do so' (ibid.: 154) that move beyond more formal analyses of ownership and law. The authors also highlight the need for deeper cultural understandings to back up these observations. This work is peculiarly suited to a bottom-up study of benefit-sharing, as it trains the focus on access to benefits of different kinds. It also echoes another central theme of the literature review presented in the previous section, namely power. Power in this theory is seen as inherent to the relationships that delineate control over access, as well as to the mechanisms that support or allow access to take place (ibid.). Specifically, the authors note that power may lie not only in explicit sites such as the law or policy, but also in self-discipline, which leads individuals or groups to act in certain ways and expresses the conditioning effects of discourse.

The authors suggest a number of categories to analyze access. First, they advise attention to 'access control', which concerns the mediation of access, then to 'access maintenance', which concerns keeping access open (ibid.: 158–159). Identifying the actors or institutions where these types of power reside helps to specify 'webs and bundles of power' where the 'strands' are the 'means, processes, and relations by which actors are enabled to gain, control, and maintain access to resources' (ibid.: 159–160). As mentioned, the theory also leads scholars to pay attention to access beyond formal legal regimes to consider access based on custom or convention. In addition, paying attention to mediation and support for access is advised, for example through capital, labor, markets, technology, authority, social identity, or knowledge (ibid.: 164–171). All of these aspects are taken into consideration in the comparison of the local case studies and aided in identifying a number of common themes. These common themes feed into research on the international level. The methodology and methods used are described briefly in the following subsection.

Analyzing space for local themes and participation at the international level

The methodology and methods used to approach the international level analyses are outlined in greater depth in Chapters 4 and 5 (see also Parks, 2018b; Parks and Schröder, 2018). The overview provided here is brief, but serves to underline the connections between the local-level case studies and the international analysis, which are intended to feed into one another. The aim of this part of the research is to look at whether, and if so where, there are discursive spaces available to raise the common themes that emerged from the case study comparison. The interest in discourse is motivated by the existing literature, which identifies a number of clear discourses that are argued to underpin global environmental governance, thus closing off spaces for local themes (see the subsection: Local communities and global environmental governance). Nevertheless, some literature on environmental discourses highlights scope for contestation: since our understandings of environmental problems emerge and change quickly, a number of discourses of environmental politics jostle to define these problems and their solutions. This provides an opportunity for different discourses to take root, and for complementary discourses to coexist (Escobar, 1998; Dryzek, 2005). Literature on the agency of indigenous peoples and local communities reports mixed findings about how these groups exercise influence in arenas of global environmental governance. Since much of this literature focuses on single COP negotiations, on specific issue areas, or on treaty texts, the analysis carried out here seeks to capture how discourses change over time, and whether they leave spaces for the themes that arose from the case study comparison. This is the first step in the international analysis, followed by a second that explores paths for participation in global environmental governance in a more practical view.

The approach for the discourse analysis is based on the picture of global environmental governance painted in the critical literature. The discourses argued to underpin global environmental governance – on capitalist or market-based (and colonial) approaches and the nature/culture divide – are operationalized and supplemented with themes that form some of the basic assumptions of the discourse approach, and are equally motivated by findings in the empirical literature. Specifically, these are about the exclusion of minority voices and the imposition of decisions. These discourse 'categories' are then joined by their opposites, which are based on the findings of the literature that is more optimistic about discursive spaces for local themes.

These categories can be considered more desirable for building discursive spaces for local communities (see Chapter 4: Analyzing the discourses of the Convention on Biological Diversity, for a detailed overview).

As already mentioned, the scope of the analysis is limited to the CBD to allow an in-depth discourse analysis over time. The CBD is an ideal candidate for a number of reasons. First, it is seen as one of the most open arenas of global environmental governance to non-state actors, and as a forerunner in this area for other arenas (Affolder, 2017). In that sense, understanding the discursive layout of the CBD can be key to getting a grasp on the best-case scenario in global environmental governance. Second, the CBD is also noted as a comparatively open arena of global environmental governance for indigenous peoples and local communities in particular, which are recognized in the treaty text along with the intrinsic value of nature (Reimerson, 2013). Third, the CBD is the most developed international legal source on benefit-sharing via its Nagoya Protocol. To analyze how the CBD's discursive shape changes over time, the texts of its COP decisions are considered. COP decisions are used to implement the treaty in concrete terms, and have been argued to move the CBD beyond the strict text of its treaty towards a dynamic interpretation of the law (Morgera and Tsioumani, 2011). This sits well with a constructivist view of norms and discourses as built in social interaction over time. In addition, COP decisions are adopted by consensus by all of the 196 parties to the CBD. They are thus the final product of discursive contestation, and should reflect the ebb and flow of different discourses over time. The analysis covers all COP decisions from the first meeting in 1994, to the fourteenth held in 2018. The texts of the decisions are sampled to yield a database of texts relevant to indigenous peoples and local communities (since the CBD groups these together in its language).[13] Each text – a paragraph or sub-paragraph – is 'coded', or assigned one or more of the discourse categories, after careful reading and discussion among researchers where required. The results build a map of the discourses relevant to local communities within the CBD over time (see Chapter 4).

The second part of the international-level analysis also focuses on the CBD. This analysis seeks to provide information about how local communities might exercise agency in addition to activities carried out directly at COP negotiations, which are already explored in existing literature (see subsection: Local communities and global environmental governance). The analysis is made up of two parts. The first focuses on how the CBD itself describes the participation of indigenous peoples and local communities in COP decisions, and captures how this changes over time. The data are drawn from the discourse analyses presented in Chapter 4, which uncover a discourse around the idea of participation in the CBD. The texts that refer to the participation of indigenous peoples and local communities in some way are analyzed in an inductive approach to examine how participation is framed. The texts referring to participation are used to formulate more abstract descriptions of participation, a process repeated until all of the texts have been classified. In addition, the texts referring to participation are coded according to where they envisage participation is to take place – at the local, domestic or international levels, at no specified level, or a combination of these (see Chapter 5: The view from above – participation and local communities in the CBD).

The final part of the empirical analysis on international participation is more reflective, and concerns community protocols. The motivation for this is the attention that community protocols have received not only in the scholarly literature (Jonas, Bavikatte and Shrumm, 2010; Tobin, 2013; Bavikatte, 2014; Bavikatte, Robinson and Oliva, 2015; Delgado, 2016) but also from policymakers (Jukic and Collings, 2013), not to mention local communities themselves. Community protocols can be seen as bridges from local to international arenas because they 'seek to help uphold the rights of indigenous peoples and other communities by filling a space at the nexus of international, national and customary law and policy' (Parks, 2018: 88; see also Bavikatte, Robinson and Oliva, 2015). By filling this space, community protocols can help local groups to defend their natural resources and access, as well as to make claims by outlining their visions of the future and definitions of development (Jonas, Bavikatte and Shrumm, 2010). Existing work on community protocols focuses on their use as legal tools, but they are also interesting from a point of view of investigating how social actors interact. As de Sousa Santos (2014) argues, existing institutions view the knowledge held by indigenous peoples and local communities as superstition, and thus irrelevant, which necessitates 'intercultural translation' that community protocols can contribute to. In addition, community protocols are interesting in that they were recognized by the CBD in its Nagoya Protocol following the efforts of NGOs and representatives of indigenous peoples and local communities amongst others (Bavikatte, Robinson and Oliva, 2015). No formal method is used to guide the reflections on community protocols. Rather, the aim is to present a systematic discussion that covers the different ways in which they might challenge the hegemonic discourses of global environmental governance via what might be termed 'countervailing power' (Fung and Wright, 2003), and how their use as conduits for international participation could look in more concrete terms.

Overview of the book

This chapter provided a general introduction to the book and its aims, discussed relevant literature, and outlined the methodology and methods used. Chapter 2 provides more detailed overviews of each of the five local community case studies. Each case study is introduced with a short *vignette*, which seeks to give some insight into the experience of arriving in an unknown community for the first time, and the impressions of the researcher. They serve to underline that the case studies do not claim any deep or complete understandings of the experiences of the communities concerned. Rather, they are attempts at conveying a general view of how communities frame issues that they consider as linked to benefit-sharing. The descriptions of the case studies focus on governance contexts and the discussions of local communities, and seek to give the reader a key to how each case fits into the thematic discussions in Chapter 3. The five case studies are on traditional goat pastoralists in Ikaria, Greece, and themes linked to access to land; traditional health practitioners in Bushbuckridge, South Africa, and themes linked to access to medicinal plants; Khwe residents of Bwabwata National Park, Namibia, and themes linked to wildlife management; residents around the Olaroz and Salinas Grandes salt flats, Argentina, and themes linked to lithium mining; and, finally, the Kelabit community of Bario, Malaysia, and themes linked to rice agriculture.

Chapter 3 expands on the case studies and presents the common themes that emerged from their comparison. In line with the exploratory approach adopted, where the case studies were approached using methods chosen to allow communities' unfiltered views linked to benefit-sharing to emerge, not all of the common themes refer directly to benefit-sharing in the terms used to define it in different international sites. The concerns raised by local communities were more often linked to the conditions that need to be in place in order for discussions about benefit-sharing to begin, and for any eventual benefit-sharing to stand a chance of being fair and equitable. The central common theme identified is recognition, which is discussed in terms of recognition from without (concerning external actors) and recognition from within (concerning intra-community dynamics). All of the other common themes that arose from the case comparison are linked to this central idea of recognition. The themes of support from non-state actors and the organization of communities are, alongside forms of community recognition, understood as necessary for communities to pursue their goals (for example in negotiations for benefit-sharing). Support from non-state actors raises some questions about how support is provided, and the limitations of project-based research, while hierarchical organizational forms are noted as best suited for communities to access authorities. However, a hierarchical organizational model may not sit well with a community's traditions.

Another common theme, albeit known from the outset, is complex governance backdrops. This is discussed as a way of understanding the ranges of external actors that communities seek to establish relationships with. Related to this point is the common theme of procedural guarantees. This is discussed with particular reference to information and voice, both of which are viewed as necessary for local communities to function in complex contexts. Finally, framings of benefit-sharing itself emerge from the comparison. First, communities underlined the importance of non-monetary benefits as well as monetary ones. Second, they underlined that benefit-sharing is a process, not a single event or contract that remains fixed over time. It should instead be understood as an ongoing relationship for building meaningful dialogue.

Although these common themes are based on just five case studies, the design of the research and their similarity to findings from other studies (in particular Wynberg and Hauck, 2014) means they are at the very least an interesting basis for an investigation of the international level. Chapter 4 presents a detailed analysis of the CBD's COP decisions to discuss whether there are discursive spaces available for the common themes that emerge from the case studies. The analysis covers CBD COPs 1–14. Based on the view of the discursive underpinnings of global environmental governance detailed in the critical literature reviewed in Chapter 1 (Local communities and global environmental governance), the analysis builds up a discursive map of the CBD over time. The chapter then reflects on the spaces available for local themes, and the discourses that deny spaces. It includes discussions of how discourses have evolved over time, and their correspondence with the different cross-cutting issues used by the CBD to organize its work. Overall, the analysis finds that there is some space for local themes, in particular since COP 7, and within the cross-cutting issue of traditional knowledge, innovations and practices, based on article 8(j) of the CBD. Participation and the recognition of traditional knowledge and customary sustainable use emerge as the broadest discursive bases for local themes. The chapter ends with a brief consideration of the COP decisions as a whole, and discusses whether a specific discourse about 'indigenous peoples and local communities' exists compared to other actors, notably industry groups. Although this research is at a preliminary stage, the evidence points to a unique discourse about indigenous peoples and local communities within the CBD.

Chapter 5 contains a more practical discussion of how local communities can exercise agency in global environmental governance by participating in the work of the CBD. Based on those texts within COP decisions that refer to the participation of local communities and indigenous peoples, an inductive analysis of how CBD COP decisions describe participation is presented. The analysis also records where participation is envisaged to take place – at the local, domestic, or international levels (or some combination of these). One of the main findings from this analysis is that the clearest paths for local communities to participate directly in the CBD are via providing inputs to international processes, and taking part in implementation. The inputs path is particularly interesting, and forms part of the motivation for the discussion in the second half of the chapter. This focuses on the potential of community protocols as tools for local communities to participate directly – or indirectly via the efforts of NGOs – at the international level. To assess this potential, the discussion asks whether community protocols can address the dominant discourses argued to underpin arenas of global environmental governance such as the CBD. Where community protocols are the result of deliberative, inclusive processes within communities, it is argued that they may be useful tools for providing inputs to the international level.

Chapter 6 concludes the book. It summarizes the main findings before opening the discussion up to future research avenues by situating them in the wider literature on global civil society and deliberative democracy in global environmental governance. The findings about how local communities frame themes linked to benefit-sharing, how these themes fit into the existing discursive landscape of the CBD, and in particular how local communities can participate in international processes, are discussed in light of literature that suggests practical models for democratic global environmental governance. The normative case for the democratic governance of the environment is based on the idea that more participatory processes ultimately lead to more efficient and just global policies, which are more sensitive to local needs and can be framed in different ways to fit local contexts. Although further research is needed to understand when and how local communities can contribute to global environmental governance, the findings of the book suggest that such outcomes can be possible, most likely through the involvement of NGOs committed to community empowerment.

Notes

1 'Benefit-sharing for an equitable transition to the green economy – the role of law (BeneLex)'. European Research Council Starting grant 335592 (November 2013–October 2018) led by Elisa Morgera. www.strath.ac.uk/research/strathclydecentreenvironmentallawgovernance/ourwork/research/benelexproject/.

2 Intra-State benefit-sharing can take place between governments and communities, or between private companies and communities (Morgera, 2016).

3 The section draws on Morgera and Parks (2014), discussed further in Parks and Morgera (2015).

4 To draw another legal parallel, the internalization stage can be equated with the idea of customary law.

5 Discussing Africa, for example, Jones (2006) notes that colonial views saw 'a decline from a prior state of pristine wilderness' (484), thus overlooking the role of humans in the definition of land.

6 The first paragraph of the preamble to the CBD states that Parties are conscious 'of the intrinsic value of biological diversity and the ecological, genetic, social, economic, scientific, educational, cultural, recreational and aesthetic values of biological diversity and its components'.

7 The formulation used in multilateral environmental treaties, and the CBD in particular. See Chapter 4: Discourse categories that challenge the critical literature, for further discussion.

8 As the author explains, these are 'a subset of third generation rights, biocultural rights have elements of group rights but differ from the latter in their explicit commitment to conservation and the sustainable use of biodiversity' (Bavikatte, 2014: 29).

9 The Rio treaties were agreed at the 1992 UN Conference on Environment and Development (also known as the Earth Summit) held in Rio de Janeiro. They are the United Nations Framework Convention on Climate Change (UNFCCC),

the UN Convention on Biological Diversity (CBD), and the UN Convention to Combat Desertification (UNCCD).

10 Reducing emissions from deforestation and forest degradation, conservation of existing forest carbon stocks, sustainable forest management, and enhancement of forest carbon stocks.

11 The methodology, methods, and ethics questions described briefly here are explored in more depth in Parks and Morgera (2019) and Morgera, Parks and Schröder (2019).

12 Respectively, https://naturaljustice.org/ (accessed 17 January 2019); https://ikariandocs.wordpress.com/about-ikarian-documentationresearch-and-action-center/ (accessed 23 July 2018); https://farn.org.ar/ (accessed 23 July 2018). The research was supported in particular by Cath Traynor, Gino Cocchiaro, Lesle Jansen, and Harry Jonas at Natural Justice, Maria Gaglia-Bareli and Frosini Koutsouti at the Ikarian Documentation and Research Center, and Pia Marchegiani and Ana di Pangracio at the Foundation for the Environment and Natural Resources, among others.

13 This database of texts, along with the discourse 'codes' assigned to each, is available in the online data appendix at www.routledge.com/9780367181871.

References

Adger, W. N. *et al.* (2001) 'Advancing a political ecology of global environmental discourses', *Development and Change*, 32(4), pp. 681–715.

Affolder, N. (2017) 'Non-State Actors', in Morgera, E. and Razzaque, J. (eds) *Encyclopedia of Environmental Law: Biodiversity and Nature Protection Law*. Cheltenham: Edward Elgar, pp. 387–398.

Bavikatte, K. (2014) *Stewarding The Earth. Rethinking Property and the Emergence of Biocultural Rights*. Oxford: Oxford University Press.

Bavikatte, K., Robinson, D. F. and Oliva, M. J. (2015) 'Biocultural community protocols: dialogues on the space within', *IK: Other Ways of Knowing,* 1(2), pp. 1–31.

Benford, R. D. and Snow, D. A. (2000) 'Framing processes and social movements: an overview and assessment', *Annual Review of Sociology*, 26, pp. 611–639.

Bixler, R. P. *et al.* (2015) 'The political ecology of participatory conservation: institutions and discourse', *Journal of Political Ecology,* 22(1), pp. 165–182.

Brand, U. and Vadrot, A. B. M. (2013) 'Epistemic selectivities and the valorisation of nature: the cases of the Nagoya Protocol and the intergovernmental science-policy platform of the biodiversity and ecosystem services (IPBES)', *Law, Environment and Development Journal,* 9(2), pp. 202–220.

Castree, N. (2008) 'Neoliberalising nature: the logics of deregulation and reregulation', *Environment and Planning A: Economy and Space*, 40(1), pp. 131–152.

Checkel, J. T. (2005) 'International institutions and socialization in Europe: introduction and framework', *International Organization,* 59(4), pp. 801–826.

Child, B. and Barnes, G. (2010) 'The conceptual evolution and practice of community-based natural resource management in southern Africa: past, present and future', *Environmental Conservation,* 37(3), pp. 283–295.

Choudry, A. and Kapoor, D. (2013) *NGOization: Complicity, Contradictions and Prospects*. London: Zed Books.

Comunidades Indígenas de las Salinas Grandes y Laguna de Guayatayoc de Jujuy y Salta (2015) *Kachi Yupi/Huellas de la Sal: Procedimiento de Consulta y Consentimiento Previo, Libre e Informado para las Comunidadades Indígenas de la Cuenca de Salinas Grandes y Laguna de Guayatayoc*. Available at: https://cl.boell.org/sites/default/files/protocolo_final.pdf.

Convention on Biological Diversity (2010) *The Nagoya Protocol on Access to Genetic Resources and the Fair and Equitable Sharing of Benefits Arising from their Utilization to the Convention on Biological Diversity*. CBD Decision X/1, Annex, Montreal.

Delgado, N. A. (2016) 'Community protocols as tools for resisting exclusion in global environmental governance', *Revista de Administração de Empresas*, 56(4), pp. 395–410.

DiMaggio, P. J. and Powell, W. W. (1983) 'The iron cage revisited: institutional isomorphism and collective rationality in organizational fields', *American Sociological Review*, 48(2), pp. 147–160.

Djelic, M.-L. (2004) 'Social networks and country-to-country transfer: dense and weak ties in the diffusion of knowledge', *Socio-Economic Review*, 2(3), pp. 341–370.

Djelic, M.-L. (2008) 'Sociological studies of diffusion: Is history relevant?', *Socio-Economic Review*, 6(3), pp. 538–557.

Dryzek, J. S. (2005) *The Politics of the Earth. Environmental Discourses*. Second edition. Oxford: Oxford University Press.

Eastwood, L. E. (2019) *Negotiating the Environment. Civil Society, Globalisation and the UN*. London: Routledge.

Escobar, A. (1998) 'Whose knowledge, whose nature? Biodiversity, conservation and the political ecology of social movements', *Journal of Political Ecology*, 5(53–82).

Finnemore, M. and Sikkink, K. (1998) 'International norm dynamics and political change', *International Organization*, 52(4), pp. 887–917.

Fung, A. and Wright, E. O. (2003) 'Countervailing Power in Empowered Participatory Governance', in Fung, A. and Wright, E. O. (eds) *Deepening Democracy. Institutional Innovations in Empowered Participatory Governance*. London: Verso Books, pp. 259–289.

Gamson, W. A. (1985) 'Goffman's legacy to political sociology', *Theory and Society*, 14(5), pp. 605–622.

Gilardi, F. (2012) 'Transnational Diffusion: Norms, Ideas, and Policies', in Carlsnaes, W., Simmons, B. A., and Risse-Kappen, T. (eds) *Handbook of International Relations*. Thousand Oaks, CA: Sage, pp. 453–477.

Gilbert, J. and Doyle, C. (2011) 'A New Dawn over the Land: Shedding Light on Collective Ownership and Consent', in Allen, S. and Xanthaki, A. (eds) *Reflections on the UN Declaration on the Rights of Indigenous Peoples*. Oxford: Hart, pp. 289–328.

Habermas, J. (1984) *The Theory of Communicative Action, Vol. 1. Reason and the Rationalization of Society*. Boston: Beacon Press.

Habermas, J. (1987) *The Theory of Communicative Action, Vol. 2. Lifeworld and System*. Boston: Beacon Press.

Hafner-Burton, E., Victor, D. G. and Lupu, Y. (2012) 'Political science research on international law', *American Journal of International Law*, 106(1), pp. 47–97.

Hardin, G. (1968) 'The tragedy of the commons', *Science*, 132(3859), pp. 1243–1248.

Johnson, N. (2014) 'Thinking through affect: Inuit knowledge on the tundra and in global environmental politics', *Journal of Political Ecology*, 21(1), pp. 161–177.

Johnston, H. and Noakes, J. A. (2005) *Frames of Protest: Social Movements and the Framing Perspective*. Basingstoke: Palgrave Macmillan.

Jonas, H., Bavikatte, K. and Shrumm, H. (2010) 'Community protocols and access and benefit sharing', *Asian Biotechnology and Development Review*, 12(3), pp. 49–77.

Jones, B. (2010) 'The Evolution of Namibia's Communal Conservancies', in Nelson, F. (ed.) *Community Rights, Conservation and Contested Land: The Politics of Natural Resource Governance in Africa*. London: Earthscan, pp. 106–120.

Jones, S. (2006) 'A political ecology of wildlife conservation in Africa', *Review of African Political Economy*, 33(109), pp. 483–495.

Jukic, E. and Collings, N. (2013) *Community Protocols for Environmental Sustainability: A Guide for Policymakers*. Nairobi. Available at: http://wedocs.unep.org/bitstream/handle/20.500.11822/8360/-Community protocols for environmental sustainability_ a guide for policymakers- 2013 Community_Protocols_Guide_Policymakers.pdf?sequence=3&isAllowed=y.

Koot, S., Beek, W. E. A. va. and Diemer, J. (2016) 'The Khwe of Namibia: foragers between game, tourism, and politics', *Anthropos*, 111, pp. 497–511.

Madzwamuse, M. (2010) 'Adaptive or Anachronistic? Maintaining Indigenous Natural Resource Governance Systems in Northern Botswana', in Nelson, F (ed.) *Community Rights, Conservation and Contested Land: The Politics of Natural Resource Governance in Africa*. London: Earthscan, pp. 241–268.

Marchegiani, P., Morgera, E. and Parks, L. (2019) *Indigenous peoples' rights to natural resources in Argentina, the challenges of impact assessment, consent and fair and equitable benefit-sharing in cases of lithium mining*, BENELEX Working Papers. 19. Available at: https://papers.ssrn.com/sol3/papers.cfm?abstract_id=3317375.

Marion Suiseeya, K. R. (2014) 'Negotiating the Nagoya Protocol: indigenous demands for justice', *Global Environmental Politics*, 14(3), pp. 102–124.

Marion Suiseeya, K. R. and Zanotti, L. (2019) 'Making influence visible: innovating ethnography at the Paris Climate Summit', *Global Environmental Politics*, 19(2), pp. 38–60.

McCool, S. F. (2012) 'Distributing the Benefits of Nature's Bounty: A Social Justice Perspective', in *International Symposium on Managing Benefit Sharing in Changing Social Ecological Systems*. Windhoek.

Mitrotta, E. (2019) *Decentralised International Cooperation: Enhancing Conservation and Sustainable Management of Transboundary Natural Resources*. University of Trento.

Morgera, E. (2016) 'The need for an international concept of fair and equitable benefit-sharing', *European Journal of International Law*, 27(2), pp. 353–383.

Morgera, E. (2018) 'Dawn of a new day? The evolving relationship between the Convention on Biological Diversity and international human rights law', *Wake Forest Law Review*, 54, pp. 101–121.

Morgera, E. and Parks, L. (2014) *An Inter-Disciplinary Methodology for Researching Benefit-Sharing as a Norm Diffusing in Global Environmental Law*, BENELEX Working Papers, 2. Available at: https://ssrn.com/abstract=2524333.

Morgera, E., Parks, L. and Schroder, M. (2019) *Methodological Challenges of Transnational Environmental Law*, BENELEX Working Papers. 20. Available at: https://ssrn.com/abstract=3354394.

Morgera, E. and Tsioumani, E. (2011) 'Yesterday, today and tomorrow: looking afresh at the Convention on Biological Diversity', *Yearbook of International Environmental Law*, 21(1), pp. 3–40.

Moses, J. W. and Knutsen, T. L. (2012) *Ways of Knowing. Competing Methodologies in Social and Political Research*. Second edition. Basingstoke: Palgrave Macmillan.

Mulligan, S. P. (1999) 'For whose benefit? Limits to sharing in the bioprospecting "Regime"', *Environmental Politics*, 8(4), pp. 35–65.

Murphree, M. (2004) 'Communal approaches to natural resource management in Africa: from whence to where?', *Journal of International Wildlife Law and Policy*, 7, pp. 203–216.

Natarajan, U. and Khoday, K. (2014) 'Locating nature: making and unmaking international law', *Leiden Journal of International Law*, 27(3), pp. 573–593.

Nelson, F. (2010a) 'Democratizing Natural Resource Governance: Searching for Institutional Change', in Nelson, F. (ed.) *Community Rights, Conservation and Contested Land: The Politics of Natural Resource Governance in Africa*. London: Earthscan, pp. 310–333.

Nelson, F. (2010b) 'Introduction: The Politics of Natural Resource Governance in Africa', in Nelson, F. (ed.) *Community Rights, Conservation and Contested Land: The Politics of Natural Resource Governance in Africa*. London: Earthscan, pp. 3–31.

Ngoitiko, M. *et al.* (2010) 'Pastoral Activists: Negotiating Power Imbalances in the Tanzanian Serengeti', in Nelson, F. (ed.) *Community Rights, Conservation and Contested Land: The Politics of Natural Resource Governance in Africa*. London: Earthscan, pp. 269–289.

Oguamanam, C. (2012) *Intellectual Property in Global Governance: A Development Question*. London: Routledge.

Ostrom, E. (1990) *Governing the Commons: The Evolution of Institutions for Collective Action*. Cambridge, UK: Cambridge University Press.

Parks, L. (2018a) 'Challenging power from the bottom up? Community protocols, benefit-sharing, and the challenge of dominant discourses', *Geoforum*, 88, pp. 87–95. doi: 10.1016/j.geoforum.2017.11.011.

Parks, L. (2018b) 'Spaces for local voices? A discourse analysis of the decisions of the Convention on Biological Diversity', *Journal of Human Rights and the Environment*, 9(2), pp. 141–170.

Parks, L. and Morgera, E. (2015) 'The need for an interdisciplinary approach to norm diffusion: the case of fair and equitable benefit-sharing', *Review of European, Comparative and International Environmental Law*, 24(3), pp. 353–367.

Parks, L. and Morgera, E. (forthcoming 2019) 'Research note: reflections on methods from an interdisciplinary research project in global environmental law', *Transnational Environmental Law*, 8(3), pp. 1–14.

Parks, L. and Schröder, M. (2018) 'What we talk about when we talk about "local" participation in international biodiversity law: the changing scope of Indigenous peoples and local communities' participation under the Convention on Biological Diversity', *Partecipazione e Conflitto*, 11(3). doi: 10.1285/i20356609v11i3p743.

Parry, B. (2004) *Trading the Genome: Investigating the Commodification of Bioinformation*. New York: Columbia University Press.

della Porta, D. and Kriesi, H. (2009) 'Social Movements in a Globalizing World: An Introduction.', in della Porta, D., Kriesi, H., and Rucht, D. (eds) *Social Movements in a Globalizing World*. Basingstoke: Palgrave Macmillan, pp. 3–22.

Reimerson, E. (2013) 'Between nature and culture: exploring space for indigenous agency in the Convention on Biological Diversity', *Environmental Politics*, 22(6), pp. 992–1009.

Ribot, J. C. and Peluso, N. L. (2003) 'A theory of access', *Rural Sociology*, 68(2), pp. 153–181.

Rihoy, L. and Maguranyanga, B. (2010) 'The Politics of Community-Based Natural Resource Management in Botswana', in Nelson, F. (ed.) *Community Rights, Conservation and Contested Land: The Politics of Natural Resource Governance in Africa*. London: Earthscan, pp. 55–78.

Robbins, P. (2012) *Political Ecology*. Second edition. Oxford: Wiley-Blackwell.

Robinson, D. F. (2010) *Confronting Biopiracy: Challenges, Cases and International Debates*. London: Earthscan.

Robinson, D. F. (2015) *Biodiversity, Access and Benefit-Sharing. Global Case Studies*. London: Routledge.

Said, E. (1978) *Orientalism*. New York: Pantheon Books.

Sarfaty, G. A. (2007) 'International norm diffusion in the Pimicikamak Cree Nation: a model of legal mediation', *Harvard International Law Journal*, 48(2), pp. 441–482.

Savaresi, A. and Bouwer, K. (2018) 'Equity and Justice in Climate Change Law and Policy: A Role for Benefit-sharing', in Jafry, T. (ed.) *Routledge Handbook on Climate Justice*. London: Routledge.

Shiva, V. (1997) *Biopiracy: The Plunder of Nature and Knowledge*. Cambridge, MA: South End Press.

Sibuye, R. *et al.* (2012) 'The Bushbuckridge BCP: traditional healers organise for ABS in South Africa', Special issue: Biodiversity and Culture: Exploring Community Protocols, Rights and Consent, *Participatory Learning and Action*, 65. Available at: http://pubs.iied.org/pdfs/G03403.pdf (accessed 9 July 2019).

Snow, D. A. (2004) 'Framing Processes, Ideology, and Discursive Fields', in Snow, D. A., Soule, S. A., and Kriesi, H. (eds) *The Blackwell Companion to Social Movements*. Oxford: Blackwell, pp. 380–412.

Snow, D. A. and Benford, R. D. (2009) 'Alternative Types of Cross-National Diffusion in the Social Movement Arena', in della Porta, D., Kriesi, H., and Rucht, D. (eds) *Social Movements in a Globalizing World*. Basingstoke: Palgrave Macmillan, pp. 23–39.

de Sousa Santos, B. (2014) *Epistemologies of the South: Justice Against Epistemicide*. London: Routledge.

Stevenson, H. (2018) *Global Environmental Politics. Problems, Policy and Practice*. Cambridge, UK: Cambridge University Press.

Strang, D. and Meyer, J. W. (1993) 'Institutional conditions for diffusion', *Theory and Society*, 22(4), pp. 487–511.

Strang, D. and Soule, S. A. (1998) 'Diffusion in organizations and social movements: from hybrid corn to poison pills', *Annual Review of Sociology*, 24, pp. 265–290.

Svarstad, H. (2004) 'A Global Political Ecology of Bioprospecting', in Paulson, S. and Gezon, L. L. (eds) *Political Ecology Across Spaces, Scales and Social Groups*. New Brunswick, NJ: Rutgers University Press, pp. 239–256.

Swemmer, L., Grant, R., Annecke, W., and Freitag-Ronaldson, S. (2015) 'Toward more effective benefit-sharing in South African national parks', *Society & Natural Resources*, 28(1), pp. 4–20.

Tobin, B. (2013) 'Bridging the Nagoya compliance gap: the fundamental role of customary law in protection of indigenous peoples' resource and knowledge rights', *Law, Environment and Development Journal*, 9(13), pp. 142–162.

Towns, A. E. (2012) 'Norms and social hierarchies: understanding international policy diffusion "from below"', *International Organization*, 66(2), pp. 179–209. doi: 10.1017/S0020818312000045.

Traditional Health Practitioners of Bushbuckridge (2010) *Biocultural Protocol of the Traditional Health Practitioners of Bushbuckridge.* Available at: http://community-protocols.org/wp-content/uploads/documents/South_Africa-Bushbuckridge_Biocultural_Protocol.pdf.

Tsioumani, E. (2018) 'Beyond Access and Benefit-Sharing: lessons from the emergence and application of the principle of fair and equitable benefit-sharing in agrobiodiversity governance', in Girard, F. and Frison, C. (eds) *The Commons, Plant Breeding and Agricultural Research: Challenges for Food Security, Agrobiodiversity and Law.* London: Routledge, pp. 41–60.

Twining, W. (2004) 'Diffusion of law: a global perspective', *Journal of Legal Pluralism,* 36(49), pp. 1–45.

Twining, W. (2005) 'Social science and diffusion of law', *Journal of Law and Society,* 32(2), pp. 203–240.

Tysiachniouk, M. and Olimpieva, I. (2019) 'Caught between traditional ways of life and economic development: interactions between Indigenous peoples and an oil company in Numto Nature Park', *Arctic Review on Law and Politics,* 10, pp. 56–78.

Uggla, Y. (2010) 'What is this thing called "natural"? The nature–culture divide in climate change and biodiversity policy', *Journal of Political Ecology,* 17(1), pp. 79–91.

Vermeylen, S. (2007) 'Contextualizing "fair" and "equitable": The San's reflections on the Hoodia benefit-sharing agreement', *Local Environment,* 12(4), pp. 423–436.

Vermeylen, S. (2017) 'Materiality and the Ontological Turn in the Anthropocene: Establishing a Dialogue between Law, Anthropology and Eco-philosophy', in Kotzé, L. (ed.) *Law and Governance for the Anthropocene.* Oxford: Hart.

Walker, N. (2014) *Intimations of Global Law.* Cambridge, UK: Cambridge University Press.

Westbrook, D. A. (2006) 'Theorizing the diffusion of law: conceptual difficulties, unstable imaginations, and the effort to think gracefully nonetheless', *Harvard International Law Journal,* 47(2), pp. 489–505.

Wilkinson, K. (2014) 'Payment for "ecosystem services" and the "green economy": green-washing or something new?', *Journal of Human Rights and the Environment,* 5(2), pp. 168–191.

Wynberg, R. and Hauck, M. (eds) (2014) *Sharing Benefits from the Coast. Rights, Resources and Livelihoods.* Cape Town: University of Cape Town Press.

2 Five views of benefit-sharing from the local level

This chapter presents overviews of five local communities discussing themes relevant to benefit-sharing. While the cases described are exploratory, and in no way claim to be representative or exhaustive, they do provide ideas about what the core concerns or themes of interest for local communities are when discussing benefit-sharing. These core themes guide the rest of the research presented in the book, and are discussed in detail in Chapter 3. The present chapter provides more holistic overviews to underpin that discussion.

The cases focus on communities in Greece, Argentina, Namibia, South Africa, and Malaysia. They were chosen to provide as diverse a view as possible of local community discussions linked to perceptions of benefit-sharing, and vary in terms of geographical and governance contexts, as well as the natural resources concerned. All the cases share a common issue – the communities concerned have at some point considered the possibility of drafting a community protocol in an effort to gain a voice in the regulation of access to or control over some natural resource they traditionally held. Discussions about community protocols were thus considered as a proxy for the presence of discussions linked to benefit-sharing (see Chapter 1: Comparative local case studies). The stages reached in these discussions vary however. In the Greece and Malaysia cases, discussions about community protocols are at very early stages. In the Argentina case, a community protocol has been drafted and publicly launched, while in the Namibia case a protocol has been drafted but not yet finalized and launched. In the South Africa case, a community protocol has been in place for some years, and was recently updated to reflect changing needs and priorities.

Each case study was approached using different combinations of methods according to local specificities (see Chapter 1: Comparative local case studies, for a detailed discussion on methodology). Each of the communities were visited twice at a distance of about two years by two researchers (the author and a legal scholar) with the exception of Namibia, in order to reconnect with the communities or those working with them and gain a longer term perspective on how the issues they were facing had changed. In each case, the researchers worked with members of NGOs engaged in long-term projects with the communities concerned. The methods used were uniformly qualitative, and included participant observation and unstructured and semi-structured interviews with small groups and individuals both within the communities concerned and at relevant local, regional, and national levels.

The chapter provides an overview of each of the five cases in turn. At the start of each section a *vignette* provides a small sketch that seeks to convey the first and most prominent impressions from each area. Some relevant information on the history of each area and relevant governance backgrounds is then provided to situate the overviews of community discussions linked to benefit-sharing. While each of the cases is unique, some high-level themes arise that are – in different ways – common to all. These are introduced at the end of the chapter. The space devoted to each of the cases varies, as some had seen more community discussions around benefit-sharing themes compared to others.

Ikaria, Greece

Flying to the island of Ikaria can be a daunting experience. The island's airport is on its windier side, and its geography is driven home as the plane is buffeted from side to side, aiming for the runway sandwiched between rocky walls on the coast. The airport also offers a first glimpse of the closeness among Ikaria's inhabitants. A policeman picking up his colleague, the greetings between everyone in the arrivals hall and everyone in departures – the lady who runs the car hire telling you to leave the car by the airport with the keys in the ignition when you leave – everyone knows it's one of hers. A short walk from the airport and all semblance of hustle and bustle melts away, replaced with a scrubby landscape perfumed with wild oregano and sage, and a calm accompanied by the sound of the sea and the odd bleating goat. The island is mostly mountainous, with stony beaches rising quickly into steep climbs. Three small towns are the sum of its urban centers, divided by winding mountain roads that take hours to navigate. In between, goats graze, olives and other crops grow, and beehives are kept.

This case study focuses on traditional goat pastoralists on the island of Ikaria, on their access to grazing land, and markets for goat meat – both of which are affected by a myriad of other issues including the effects of climate change and Greece's membership of the European Union (EU). Traditional Ikarian pastoralists feel that their very occupation is under threat and likely to disappear in the face of these apparently insurmountable challenges. The research in Ikaria took place over two visits in 2014 and 2017, and included interviews and conversations with a range of actors – pastoralists, representatives of local authorities, veterinarians, and environmental NGOs.

Ikaria is a small and fairly sparsely populated island of about 255 square kilometers in the northeast Aegean island group. It is mountainous and largely covered with forest – a challenging climate for agriculture. Ikaria's economy remains relatively traditional compared to more touristic Greek islands, and pastoralism is a key traditional occupation. Politically, Ikaria has a strong communist tradition (popularly known in Greece as 'the red rock') linked to the exile of thousands of communists to the island during and after the Greek civil war (1946–1949) (Papalas, 2005). In recent times the island has gained some notoriety as a 'blue zone' due to its residents' longevity, linked by many to the healthy local diet as well as its culture:[1] Ikaria has a strong tradition of self-organization, including the use of land, solidarity, and community ties, expressed both through everyday activities and regular community celebrations or *paniyiri* (Bareli-Gaglia, 2010). The following provides an overview of the governance backdrop in Ikaria to situate the discussions linked to benefit-sharing and the issues raised among traditional pastoralists.

A key component in the governance backdrop for the traditional pastoralists is Greece's membership of the European Union. Greece joined what was then known as the European Community in 1981 following the end of the military regime in 1974. The EU today is a powerful regional bloc comprising 28 member states (though the UK is, at the time of writing, scheduled to leave the organization). The organization began as a mechanism for the international administration of coal and steel following the Second World War, but developed over time into a much wider reaching economic and political organization. As the organization developed, competences for regulating a range of different areas have passed from its member states to the supranational and intergovernmental EU institutions. Simplifying, the heads of State and government of the member states of the EU request supranational actors in the European Commission to develop rules and regulations for areas of EU competence that are then binding on all member states. The legislative decision-making process for these directly applicable rules has changed over time, but at present involves the intergovernmental Council of the EU, where the relevant State ministers for the subject matter at hand gather, and the direct representatives of the EU's citizens who sit in the European Parliament, as co-legislators.

This legislative process applies for directives, where member states are free to choose the method of application so long as the objectives of the directive are fulfilled, and regulations, which must be applied directly in member states. Some policy areas, including the Common Agricultural Policy (CAP), have historically been implemented in more supranational ways via the European Commission. This is still the case, though the margin of flexibility for member states has been increased. Membership of the EU thus implies that a country adopts all of the decisions taken by the EU and implements them within its borders, and faces sanctions from the body's Court of Justice if it fails to do so. Though the EU is not a full federation, membership does imply accepting decisions that may not be seen as in the interests of a member state. In addition, as a primarily economic organization, the EU has also adopted a single currency (the Euro). Though not all member states participate in the currency, Greece is a member of this group, and as a consequence must also comply with fiscal criteria set at the supranational level.

Different EU rules and regulations have clear and direct consequences relevant both for Greek national politics and for the pastoralists in Ikaria. Two major policy areas, as well as the EU's approach to the management of the financial crisis in Greece, stand out in this vein. The first is the Common Agricultural Policy and its systems of subsidies. The CAP was one of the first major policy areas in the EU, as preoccupations about food security and modernizing the European agricultural sector were prominent at the organization's beginnings in the late 1950s. The main objectives of the CAP, which have remained rather stable over time, are to increase agricultural productivity, ensure fair living standards for farmers and stabilize agricultural markets, to guarantee food supplies, and to ensure reasonable prices for consumers. The methods for achieving these objectives included the introduction of minimum prices and harmonized price supports, common protection for European farmers from external competition, and funds for market interventions and structural improvements. These methods brought unintended consequences however – fixed minimum prices led to overproduction and increasing inequalities between small-scale and industrial farmers as well as environmental problems, amongst others (see e.g. Dinan, 2010).

To attempt to address these problems, the 1992 MacSharry reform of the CAP sought to move methods away from price supports, introducing direct payments to farmers and owners of livestock instead. In Ikaria, many residents took on extra goats or acquired animals in order to take advantage of the subsidies the reform had made available by linking them to animal numbers. In 2003 another reform – known as the Fischler reform – sought to deal with new problems that had arisen from linking subsidies with animal numbers alone, including overpopulation. This reform thus began a move towards 'cross-compliance', where subsidies were linked to a range of factors including animal welfare and environmental standards (ibid.). In Ikaria, this reform led to those animals that were no longer regarded as profitable being released, feeding into a progressive increase in the island's goat population as well as overgrazing. In 2017, in the context of a financial crisis in Europe that hit Greece particularly hard and saw various fiscal measures and cuts imposed in the country, moves to ensure CAP rules were applied more thoroughly and strictly were noted in Ikaria. These included cross-compliance requirements for subsidies, such as welfare and environmental rules about barns, drainage, and waste disposal systems. Difficulties in compliance, linked to the particular circumstances of Ikaria (explored in the following subsection), meant that pastoralists risked losing subsidies and their livelihoods, and a local committee was in the process of being created to attempt to prevent this (see Gaglia Bareli *et al.*, forthcoming 2020.).

A second area of EU policy with clear ramifications for pastoralists is food safety (and by extension animal welfare). One fundamental obstacle to reducing the goat population in 2015, for example, was that Ikaria has no approved slaughterhouse, as goats had been killed using traditional methods. EU legislation, stemming from the various food safety scares and crises of the 1990s, provides that all animals whose meat is to be sold must be killed in line with various rules and in slaughterhouses that meet certain criteria. Goats from Ikaria must thus be transported to Athens for slaughter rather than killed locally using traditional methods. The transport of live animals has created prohibitive costs however, as EU animal welfare rules applying to the care of animals being transported, the specifications of the vehicles used, and the skills required of drivers, must be observed. The slaughterhouse issue also maps onto local politics, as pastoralists felt that municipal authorities were partly responsible for the failure to approve the construction of a facility on the island.

One theme that emerges as central in the story of the local application of EU rules in Ikaria is a lack of clear information and understanding. Across different waves of reforms, the CAP and rules for subsidies became steadily more opaque, as did the reasoning behind the need for a slaughterhouse. This translated into a general feeling of exclusion vis-à-vis the EU among pastoralists, but also to some extent from local authorities. Doubt and uncertainty also surrounded a land use map and added to the sense of exclusion pastoralists felt with regard to local, national, and European decision-making more generally. This map, approved by the municipal authority that left office shortly before the first research visit in 2014, continued to be an unresolved issue during the second visit in 2017. It was generally considered to be inaccurate, partly as a result of its being adopted without proper involvement from local groups. In addition, there was no clarity about the legal significance and consequences of the map, and pastoralists feared that the designation of some land as private would lead to their exclusion despite a long history of traditional access and a generally shared understanding of the land as commons. Finally, the map was rumored to be linked to a plan to install wind turbines, as substantial tracts of land were designated for industrial use. This plan was attributed to the national government's need to generate income in the context of commitments seen as imposed by the EU in return for financial aid to overcome the sovereign debt crisis.

Traditional pastoralists in Ikaria

To provide an idea of the issues that Ikarian pastoralists face, their organization, and their understandings of benefits, a step back in time is necessary. When pastoralists and others spoke about these themes they often contrasted present difficulties with the past. In the past, pastoralists were key to local food security and health – goat meat was a particularly healthy food source, bred on the island from a local and semi-wild population (which remains the case today). With no natural predators, goats roamed freely – hence the perplexity and lack of readiness to provide barns in line with CAP rules – and their natural diet resulted in healthy, lean meat. To solve problems in the coexistence of freely roaming goats and agriculture, villages took communal decisions to build and maintain a system of walls and gates, keeping goats away from crops. However, this system became obsolete when roads were built, and in the more recent past the role of field guards (or agricultural field officers – State employees) was underlined as key to keeping the goat population at a sustainable level. Historically, pastoralists obtained access to land for grazing through communal village agreements that amounted to customary access to land owned under different property regimes.

Pastoralists described how they felt more respected and valued in the past. For example, as the only local community members to regularly visit the highest and more inaccessible parts of the mountains, they located and maintained freshwater sources. Their extensive knowledge of the island's terrain also contributed to the prevention and tackling of forest fires, a common danger in the hotter months, through strategic grazing to clear undergrowth, and through alerting and guiding firefighters. To avoid overgrazing, pastoralists moved animals according to the different seasons, sticking to lower ground in colder seasons and higher ground in the hotter ones. This brief sketch of pastoralists as they saw themselves in the past underlines a sense of their importance and of their contribution to Ikaria's social and political fabric.

The picture painted of the situation facing pastoralists in 2014 was one where these balanced relationships had been overturned. The central problem with regard to pastoralists, as raised by local government actors, environmentalists, pastoralists themselves, and other residents, was the rising number of goats. Numbers were universally acknowledged to be too high, and overgrazing a growing problem. The overall situation was described as follows. The initial increase in goat numbers was generally linked to the CAP reforms discussed earlier, then exacerbated by the lack of natural predators and by shrinking possibilities for legal slaughter and the sale of meat. While derogations to these rules allowed for some traditionally slaughtered meat to be sold for paniyiri, the traditional local festivals, this was not enough to solve the problem. This caused the beginnings of friction between pastoralists and other groups on the island, particularly around access to land, stoked by the above-mentioned land use designation map.

Though they acknowledged that problems existed, and showed willing to participate in various schemes to resolve them, pastoralists felt that blame was laid exclusively at their door despite the complex governance backdrop behind the population issue. Pastoralists also felt that the balances of the past had disappeared: with fewer pastoralists and too many animals, as well as tourist developments on the coast, flocks were no longer moved according to the seasons. Climate change was also raised as a contributing factor here: as winters have become milder, the need to move flocks reduced, and reduced rainfall affected the availability of grass leaving pastoralists more reliant on bought feed (Gaglia Bareli *et al.*, forthcoming 2020). In 2017, the issue of overpopulation had been eased. Nevertheless, issues related to the land use map, linked in turn to the governance context discussed earlier, remained, as did fears about the compliance conditions for CAP subsidies.

This is the context is which pastoralists had begun to discuss how to address the various issues they faced. Discussions were difficult to coordinate however, and little progress had been made. It is worth noting in this vein that no functioning pastoralists' association was in place. Though a legally registered association existed, it was not active, as the sacrifices of time and energy judged necessary to revitalize it were deemed too cumbersome, particularly given the general feeling of exclusion from different levels of decision-making experienced by pastoralists. Pastoralists received some support from a local NGO, the Documentation, Research and Action Center of Ikaria, which had, amongst other things, arranged events to explore the possibility of pastoralists drafting a community protocol. However this support had not resulted in any regular collective discussion amongst pastoralists and other residents by 2017. Overall, the Ikaria case saw the least progress in terms of collective community action to discuss how to address common issues.

Salinas Grandes, Argentina

The air is thin in the villages around the salt flats of La Puna, high in the Andes in the north of Argentina. The Salinas Grandes salt flats are blinding white in the sun as we drive past, before grass gradually reappears. There is little to see as the flats stretch into mountains, though a llama in the road stops us at one point close to a large billboard plastered with a message from a mining company. It announces the words 'honesty' and 'transparency' in towering letters to the mostly empty landscape. Heading for the biggest village in the area and giving a lift to local residents traveling home for the day from their fields, we chat and are told to move slowly, eat little, and drink plenty to head off altitude sickness. In the village, a few hostels provide accommodation for mine workers, and a few small restaurants provide food. The only other visitors we encounter are tourists passing through, looking for transport to the nearby border with Chile. The nights are cold, well below zero, and the sun can burn in a moment during the day, particularly when reflecting off the salt. Communication and travel are challenging – cellphone coverage is scarce and there is no local transport. One meeting between community representatives starts four hours late because of a lack of passing traffic for hitchhiking.

In Argentina, the case study is based on two visits in 2015 and 2017 and a number of informal conversations, group interviews, and individual interviews with community representatives, as well as semi-structured interviews with a range of local, provincial, and national authority representatives and relevant NGOs. The research focused on communities seeking different forms of engagement from mining companies interested in extracting lithium reserves in the salt flats. The main focus is a group of 33 communities from the Salinas Grandes and Laguna de Guayatalloc area. Community members were concerned about mining exploration activities (and the possible routing of a major rally) on their lands that they had not been fully consulted about. To address these issues, the 33 communities drafted a community protocol describing a culturally appropriate approach to consultation and consent. A secondary focus of the research concerned a group named the Colectivo Apacheta, composed of members from communities in the area around the Olaroz salt flat. Mining has been underway here since 2015, and expansions and fresh projects are planned for the coming years (Marchegiani, Morgera and Parks, 2019). The community members involved in the Colectivo Apacheta were also concerned about consultation, which they felt had not been secured from the community as a whole before mining began. Both of these groups had wider concerns about the effects of mining on scarce freshwater supplies in the long term – proper consultation, with clear long-term agreements about environmental protection and benefit-sharing, were seen as key to any meaningful consent for mining on indigenous lands.

A brief look at the historical, political, and legal backdrop in place is a good starting point to begin to understand the steps and decisions made by the communities discussed in this case. Complex governance backdrops are common to all the cases discussed in this chapter. However, in the Argentina case this complexity is closely and explicitly tied with communities' views of the country's history and the legitimacy of its authorities. This provides some keys to understand the communities' actions and views, and their framings of issues around benefit-sharing.

Argentina's colonial history continues to be felt strongly by its indigenous peoples today. The country gained independence from Spanish colonial power in 1816, followed by various periods of undemocratic military rule, until the present period of civilian democratic government began in 1983. Argentina is a federal State, and all powers not specifically delegated to the federal government belong to the provincial governments, including power over natural resources. This balance of power shifted in areas relevant to this case study with the Constitutional reform of 1994. First, it recognized the ethnic and cultural pre-existence of indigenous peoples and community rights to their lands.[2] This meant that provincial governments began to transfer land titles to local communities, with knock-on effects about their rights to consultation on mining projects due to the country's Mining Code, which requires landowners' permission (Anaya, 2012: 11). Second, the reform saw the addition of an article on the right to a healthy environment,[3] opening the way for a new general Environmental Protection Law (Ley General del Ambiente, Law No. 25.675) that included rules on impact assessments and consultation neglected in the Mining Code. Another layer of opportunity, but also complexity, derives from the fact that Argentina gives supremacy to international human rights law, which includes sources such as the International Labour Organization's Convention 169 on consultation and consent (International Labour Organization, 1989; Marchegiani, Morgera and Parks, 2019). Overall, the governance backdrop to these cases provides many opportunities for local communities to assert their rights, particularly through legal avenues. At the same time, however, the overlapping and sometimes contradictory content of international, federal, and provincial law, as well as problematic or patchy implementation, contributes to problems experienced in local communities' efforts to protect natural resources.

This backdrop applies to all environmental conflicts in Argentina, including this case on lithium. Lithium mining in Argentina began around 1997, and is thus a relatively recent phenomenon. Global demand has since increased, and is often linked to the attempt to move away from a fossil fuel-based global economy given that lithium is a key component of li-ion batteries, used, for example, in electric cars. Lithium is however rather difficult to extract from some sources, and the most accessible reserves are in salt flats. Of these, among the richest sources on the planet are found in the so-called 'lithium triangle': a group of salt flats located in the Andes close to the respective borders of Argentina, Bolivia, and Chile. The government of Argentina has in recent years taken steps to encourage industries linked to lithium via tax breaks for all extractive industries as well as specific support for lithium processing (see Marchegiani, Hoglund Hellgren and Gomez, 2018). In 2016, Argentina's exports and extractions of lithium were growing faster than any other country's – a fact also suggested by the numbers of lithium-mining projects either at an exploratory stage or further in the areas where the communities at the center of this case study live (ibid.).

The stories of two community groups active on questions around lithium mining were explored. Of these, the story of the Salinas Grandes communities is covered in more depth, since it includes the drafting and publication of a community protocol, which provides a wealth of information about views on authorities, mining, and consent. The description of the Colectivo Apacheta is more cursory, but provides an interesting point of comparison.

The 33 communities: Salinas Grandes and Laguna de Guayatayoc

Mining companies began exploration activities in the Salinas Grandes area in 2010, yet by 2015 no mining project had been submitted to provincial authorities. The communities living in the Salinas Grandes area were concerned by these explorations on their lands, as they had not, in their opinion, been consulted. Grouping together, the communities founded the Table of the Native Peoples of Salinas Grandes and Laguna de Guayatayoc and undertook a number of actions. Their community protocol was thus only one action among many, and seen as complementary to more formal legal actions drawing on the opportunities afforded by the complex governance backdrop of the country. Working with locally based lawyers who provide support in cases linked to indigenous peoples' rights, the group brought their worries and complaints about the lack of consultation on exploratory activities to both the national and international levels.[4]

The community protocol is of central interest for this work, as it outlines the communities' views on consent and community voice, both of which can be understood as crucial for any kind of future benefit-sharing to be possible (Comunidades Indígenas de las Salinas Grandes y Laguna de Guayatayoc de Jujuy y Salta, 2015). The process of drafting the protocol was entirely driven by the communities (through the Table of the Native Peoples of Salinas Grandes and Laguna de Guayatayoc), with support provided by locally based lawyers and by national NGO FARN only when requested. The purpose of the protocol, based on community consultations, was to inform all those external actors who wished to work in the area or with the communities of the procedures and standards they expected to be respected in consultations. The original idea had been to draft a protocol that spoke directly to issues of relevance to lithium mining, but over time a broader approach was preferred as the group identified other environmental issues where consultation was also seen as key.[5]

The protocol placed consultation in the context of the community's culture in order to design a consultation procedure that was appropriate for the communities. The protocol thus aimed to make the relevance and meaning of law, including international law and ILO Convention 169, clear to community members by linking it with systems and processes they were familiar with. For external actors, the protocol aimed to outline the importance and relevance of the community's culture and outlook – their worldview – and link it to consultation processes. Thus, rather than outlining the benefits and problems experienced by the communities – as other community protocols explored later in this chapter do – this protocol describes the history of the communities, the basis of their expectations on consultation in international, national and provincial law, and details advice on how such procedures should unfold, presenting a comparison with the salt cycle.

Argentina is described as a State run by descendants of colonial powers, who continue a history of exploitation of indigenous peoples through taxes and other more violent means. The protocol illustrates the history of struggle against colonial powers including battles, forced migration, and mining activities that saw many abandon traditional livelihoods. Oppression is thus described as a situation that needs to be addressed – this can be done, in part, through proper consultation. The consultation procedure outlined is compared to the salt cycle. The key point here is that salt is not, in these communities' worldview, an economic or nutritional good alone, but a living thing. By linking law and the consultation process to the salt cycle, the protocol not only renders the law clearer to community members, but also illustrates that consultation too is a living, dynamic process. In addition, linking these shows how the traditional knowledge of the land held by the communities is significant for consultation processes. Describing the process of an ideal consultation makes sense in a context where consultation is mandatory, but implementation is patchy and little information about the form of consultation procedures is available. The protocol thus also takes the opportunity to make it clear that the basis for any consent is the compatibility of a project with *Buen Vivir*, which is 'the process of full communal life on our land. It is being one and the same with the communities from its very roots. To achieve *Buen Vivir* means knowing how to live and thus how to live with others' (Comunidades Indígenas de las Salinas Grandes y Laguna de Guayatayoc de Jujuy y Salta, 2015).

Regarding the legal and governance backdrop on environmental issues, the protocol identifies the communities' ideas about potential problems in environmental impact assessments. It underlines worries about companies producing fair and impartial impact assessments; about community participation being unsatisfactory or omitted; and about the opinions of communities lacking any binding power. Regarding the first of these, interviewees in Argentina confirmed that neither federal nor provincial government offices have sufficient resources at their disposal to produce necessary expert information and research for impact assessments. The body that proposes a project thus often funds information of this type: in this case, mining companies produced information about water sources. The communities are suspicious that this information is biased, and the protocol underlines the need for independent information in this regard. Crucially, it also emphasizes the format of information, which must be accessible as opposed to technical reports. The protocol also calls for genuine consultation with communities by underlining the need for the proper scheduling, number, and locations of meetings. For example, meetings should be advertised in good time, be held in accessible places, and continue to be held until the communities are ready to make a decision. On the latter point, the protocol underlines that consultation should involve dialogue rather than the mere provision of information.

The final draft protocol, named Kachi Yupi (tracks in the salt) was completed in July 2015 (Comunidades Indígenas de las Salinas Grandes y Laguna de Guayatayoc de Jujuy y Salta, 2015). After its launch, timed to coincide with elections held late that year, the document was discussed by the newly elected provincial government. As a result, mining permits in place for the Salinas Grandes area were suspended, and there was a debate to pass a provincial decree which would see the protocol officially recognized. After three years, however, no decree had been adopted, and permits were reactivated in 2018 (Marchegiani, Morgera and Parks, 2019). Nevertheless, the National Ombudsman's office did recognize the protocol, and recommended that all provincial agencies follow it when consulting communities in the area. As of early 2019, none of the companies carrying out explorations in the area had shown willing to follow the terms of the protocol, and the communities began to mobilize again using more contentious methods following a fresh attempt to claim their right to consultation. Some community members now feel that all contacts should be stopped, and lithium mining rejected outright.[6]

The Colectivo Apacheta: Salar de Olaroz area

Lithium mining in the Salar de Olaroz area began in 2015, and further projects were planned at the time of writing (Marchegiani, Hoglund Hellgren and Gomez, 2018). Community permissions and a benefit-sharing agreement are in place for existing projects. Nevertheless, at the time of the first research visit in 2015 some community members from the area had continued concerns. They argued that community consultations were not inclusive or meaningful, and they were particularly worried about the effects of mining on freshwater supplies. The salt flats in this case are in the second driest area on the planet, and lithium mining uses large amounts of water. The worry was that water could run out, become polluted, or that saltwater and freshwater supplies could become mixed, threatening local livelihoods based on pastoralism and small-scale agriculture. Concerns about consultation and information were linked to this – some community members felt that information about these issues was not provided in an accessible format during community consultation processes.

The Colectivo Apacheta was formed in March 2013 by some of these concerned community members, mostly pastoralists from a variety of communities in the Olaroz salt flat area, to articulate these points. Apacheta, meaning cairn, was the name they chose to reflect their hopes about building their support over time. The group employed a lawyer who filed a series of administrative complaints against provincial institutions linked to mining, as well as the national government, echoing the legal paths followed by the Salinas Grandes group. The writs were all based on the claim that consultations during the environmental impact assessment did not fulfill national and international obligations to ensure the participation of indigenous communities. In this vein, members of the group described their concerns about consultation beyond the water issue. First, they recalled that consultations were held at regular monthly town meetings, which are held to deal with the normal business of local governance. That these counted as consultations proper for the impact assessment was not, in their opinion, clearly communicated. In addition, they felt the meetings were not fit for the task – they were not dedicated solely to discussions of the mining project, and group members felt that company representatives were unwilling to engage in real dialogue. Some members felt that once it was understood that they were opposed to the project they were no longer invited to discussion forums.

Second, the members of the group interpreted the mining company's approach as linking consent to the provision of employment benefits. This, they believed, played a role in securing passive consent to the project. Third, some in the group believed that signatures on the consent forms in one village (each village in the area was consulted separately) did not all belong to eligible community members – this is significant since that village's charter requires signatures from 50 percent plus one resident. The companies were thus felt to have wrongly interpreted the signatures as valid for consent. The fourth and final issue concerns information, as already noted. Group members cited several occasions when they had requested information from authorities with little success: answers either took a long time to arrive, or no answers were given due to confidentiality. The environmental impact assessment was available for community members to consult, but was described as too technical for a non-expert audience. Not all of the members of the communities in the Olaroz area shared this opinion of the consultation procedure, creating tensions between those who supported and those who opposed mining projects.

The worries of the Colectivo Apacheta echo those addressed in the Kachi Yupi community protocol. They underline a general view of consultation and consent as a crucial step to be taken if any eventual fair and equitable benefit-sharing is to take place, and link this to struggles for self-determination in historically oppressed and marginalized communities. This serves to underline the importance of establishing trust to achieve genuine dialogue.

Bwabwata National Park, Namibia

A busy road cuts through the part of the Bwabwata National Park we visit, the main thoroughfare through this long 'panhandle' of Namibia flanked by borders with Angola and Botswana. Elephants in the road frequently stop large trucks and cars, and accidents are common, particularly at night. The wildlife is a constant in the lives of those living in the villages within the park, and managing wildlife is an important part of their livelihoods. At campsites the proximity of wildlife is made clear – we should check our tents for snakes, and make sure we do not leave our tents after we turn in for the night: elephants and other animals are free to wander through the camp. We should avoid getting close to the river too – the crocodiles are hibernating at this time of year, but best not to risk it. In other campsites we are advised to sing to keep any local hyenas at bay. One evening, some of our hosts have to leave to attend a road accident – an elephant has been killed, and they must help to remove its tusks before poachers do so. All of the elephants in the park have bullets in them, they say.

Of all the case studies presented in this book, Namibia was the only destination visited just once. This circumstance was the result of various challenges in coordination and the postponement of the launch of the community protocol that formed the focus of the research. The visit was paid in June 2015 to coincide with a planned series of visits to Khwe[7] community villages in Bwabwata National Park. These visits were intended as an occasion for the committee charged with drafting a community protocol to present their work to fellow community members. Representatives of organizations that supported the Khwe community in drafting the protocol, as well as with other matters, accompanied them on this small tour: Natural Justice, Integrated Rural Development and Nature Conservation (IRDNC), and the Legal Assistance Centre based in the capital Windhoek. Representatives of the national Ministry for Environment and Tourism (MET) were also present.

As in other cases, the reasoning behind the drafting of a community protocol is connected not only to contemporary factors, but also to the history, law, and politics of the area – particularly sensitive due to relatively recent conflicts. The area now known as Bwabwata National Park (the name dates from 2007) is located on a strip of land, the 'panhandle' of Namibia, which comprises the Kavango and Caprivi regions.[8] Angola lies to the north, and Botswana to the south. The park's history is necessarily intertwined with that of the country as a whole. It was first designated as a nature reserve in 1937 under German colonial rule, which began in most of Namibia in 1884. After the First World War, administration was handed to South Africa, and the country shared in the injustices of apartheid rule. An ongoing struggle for independence accompanied this history, and intensified in the 1960s. During this decade the South-West Africa People's Organization (SWAPO) emerged as the leading force for independence. In the same decade, in 1969, what was now named the Caprivi Game Park[9] was declared a military zone by South Africa, and occupied by the South African Defense Force (SADF). Elite troops were stationed there to target SWAPO bases established in Angola, which had gained independence in 1975. These troops obliged the local Khwe community, whose skills in tracking were well known, to assist them. Some commentators see this as a source of residual distrust between the Khwe community and SWAPO, who have governed Namibia since its independence from South Africa was won in 1990. Tensions were exacerbated after a more recent episode in 1994, when an attempt was made by leaders in the East Caprivi region to secede from Namibia, with rumored support from a Khwe leader. Awareness of this political history is relevant to a basic understanding of the case. It is important to underline, however, that community members and others stressed the good relations between the community and the government in Windhoek, which had only improved and strengthened over time. This was often attributed to close working relationships forged through Namibia's CBNRM programs (see Jones and Weaver, 2009; Jones, 2010; Jones, Diggle and Thouless, 2015).

The role of the Khwe community in natural resource management within Bwabwata National Park was a central reason behind the decision to draft a community protocol (though the document reflects beyond this role). The community wished to document the various reasons why their presence in the park was crucial for good natural resource management, and to outline some challenges they were facing linked to the reproduction of their knowledge that could in turn affect their role as stewards of resources in the park. The challenges faced by the community are, as in other cases, linked to the governance structures in place in the country. What marks this case out as unique, however, is the fact that the community plays a central role in the Kyaramacan Association, which holds hunting and lodge concessions as well as harvesting rights that constitute a functioning benefit-sharing arrangement between the Namibian government and all of the communities residing in the park area.[10]

The community's view that their position in the management of the park and the role of the Kyaramacan Association needed to be underlined is linked to the governance backdrop in the country. The relevant framework here includes not only the regulation on national parks, but also the legal recognition of the rights of traditional communities and authorities. All Namibians hold fundamental rights guaranteed in the country's Constitution, which also recognizes the authority of customary law as long as this does not clash with Namibian law. Traditional communities are also addressed in the Traditional Authorities Act (Act No. 25 of 2000), which allocates power to recognized authorities to ensure the protection of traditional knowledge and customs, the sustainable use of natural resources, and to receive assistance and information about development projects within their jurisdictions. At the time of the research, there was no such legally recognized Khwe traditional authority, and the group's interests were represented by the Mbukushu authority, also resident in the area.[11] This was a concern for some Khwe community members, in part due to a perception that Mbukushu residents kept too many cattle, thus contributing to a depletion of natural resources in the park through overgrazing (Jones and Dieckmann, 2014).

As for the governance of national parks and natural resources more generally, Namibia is well known for its commitment and application of CBNRM policies (Jones and Weaver, 2009; Jones, 2012; Jones, Diggle and Thouless, 2015). CBNRM usually takes place in Namibia through the establishment of conservancies on communal land, but these cannot be set up in national parks, which are State-owned. Although the Kyaramacan Association in Bwabwata National Park is very similar to a conservancy, and involves the Khwe in natural resource management, the rights that flow from this involvement are thus less clear.[12] National parks also present a more complex situation regarding residents. Existing legislation on nature conservation makes no reference to residents in parks. However, when Bwabwata National Park was officially gazetted in 2007 it hosted resident communities. A policy on communities living in and near national parks was published in September 2013 (National Policy on Protected

Areas' Neighbours and Resident Communities) by the MET. This policy highlights 'equitable *access* to benefits from protected areas' and a new framework for zoning 'in order to accommodate the development needs of resident communities'. The policy also focuses on communication between the government and park residents and neighbors to achieve conservation and development goals.

Overall, this complex and contrasting governance backdrop in Namibia contributes to an understanding of the Khwe community's aim to clarify their position as residents and wildlife managers in the park. The following subsections discuss these elements as expressed in a draft community protocol, followed by a brief outline of the community's role in the Kyaramacan Association.

The protocol of the Khwe community of Bwabwata National Park

The Khwe community protocol was drafted over a period of about four years. At the time of writing (early 2019), the community is awaiting confirmation from the MET, which was closely involved in the drafting process, to launch the protocol publicly.[13] The process received its initial impetus when the Khwe community took part in a workshop alongside Natural Justice in 2012 where the possibilities offered by community protocols were discussed (see Natural Justice and the ABS Capacity Development Initiative, no date). The decision to draft a protocol was taken (with the support of Khwe community members from Botswana) in 2014 and a Custodian Committee was elected. The Custodian Committee, composed of four men and four women representative of both community elders and youth, took part in various training courses to support them in the drafting process. Training covered strategies for identifying and mapping resources, presenting customary institutions, and placing these in the context of national and international law. After the Custodian Committee drafted the protocol, assisted by a staff member from the MET, the text was presented to and discussed in the Khwe villages located in the park. It was at this point in the process that the research visit took place.

The Khwe Custodian Committee was also supported throughout the process of drafting the community protocol by three actors – IRDNC, which had a permanent staff member based in the park, Natural Justice, and the MET. Further support also came from the Legal Assistance Centre based in Windhoek and the Center for Indigenous Knowledge and Organizational Development, an NGO based in Ghana.

The content of the protocol, though it has not yet been officially launched, outlines why the Khwe consider themselves to be stewards of the park and its natural resources, and how their knowledge produces benefits. Underpinning this perception as stewards of natural resources is the importance the Khwe community attach to the land, which is both a home and a source of livelihood to be looked after. Land provides food, shelter and medicine, and wildlife in the area is considered the community's 'herds'. Therefore, both land and wildlife are seen as resources to be protected and used sustainably. The traditional methods the Khwe use in the management of fauna and flora in the area in which they live are wide-ranging. Some methods, such as sustainable harvesting and wildlife-tracking techniques, are also used in CBNRM within the park through the Kyaramacan Association, which employs game guards and natural resource monitors. It is upon this basis – as a community with an identity and knowledge closely tied to the land – that the protocol seeks to outline the Khwe's visions and priorities, as well as their organization and decision-making processes, including consultation procedures. The document also notes barriers and challenges, including those linked to the governance backdrop, and the community's rights under national law (Natural Justice and the ABS Capacity Development Initiative, no date).

Among their visions and priorities, the Khwe place a great deal of emphasis on the need for actions aimed at preserving and transmitting traditional knowledge – which is the basis of their livelihood – to future generations, in particular through setting up a dedicated school (the Traditional Environmental Knowledge Outreach Academy – TEKOA). During meetings with different stakeholders prior to the presentation of the draft protocol, members of the Custodian Committee elaborated on this point. A consequence of the different barriers they faced, including access to education and employment opportunities, was a rise in problems of alcoholism and a general disinterest among the younger generation in learning traditional knowledge and skills. The school, they argued, would be crucial to allow the preservation of traditional knowledge, equipping the next generation as well as those outside the community who wished to learn. A charge would be levied for those outside the community to attend the TEKOA, allowing the initiative to become a source of monetary benefit-sharing from traditional knowledge.

The potential of the protocol was boosted in 2017 with Namibia's Access to Biological and Genetic Resource and Associated Traditional Knowledge Act. This Act defines community protocols, and 'foresees the development of regulations' on these (Natural Justice and the ABS Capacity Development Initiative, no date: 35). Other themes covered in the Act include the protection of traditional knowledge, the recognition of customary law on traditional knowledge, and the extension of CBNRM approaches (ibid.).

The Khwe and the Kyaramacan Association

The Kyaramacan Association (KA) merits a little more discussion, as it is through their role in this body that Khwe communities partake in benefit-sharing. The KA was established in 2005 and recognized in 2006, with the aim to 'promote the general welfare, conserve and develop the resources of the residents of the Bwabwata National Park' (Constitution of the Kyaramacan Association, 2012, cited in Jones and Dieckmann, 2014: 391) and allow resident communities to benefit from tourism activities in the park (ibid.). The KA's board is elected by the residents of the park, and represents the Khwe and Mbukushu groups (though not the smaller !Xun group of residents). Its composition is proportional, and, as the largest resident group, the Khwe has more voices within the KA. Koot (2013) finds that some Khwe residents see the KA as a Khwe organization that can, to some extent, fill the role of a traditional leader.

The KA is a vehicle for benefit-sharing within the park as well as a direct employer of community game guards and female community resource monitors who use traditional knowledge in their work. The KA receives and manages income from tourism, hunting concessions, and the sale of the plant Devil's Claw, used internationally as a natural painkiller and fever remedy, which is harvested in the park. On the tourism front, the KA holds campsite and lodge concessions, secured in 2009 after a long and complex history of competing claims between Khwe and Mbukushu actors as well as private investors (Koot, 2013). Three community campsites were already being run by the KA at the time of the research visit in 2015 according to the IRDNC, while another campsite and a lodge were in development. The KA also has two craft outlets. All of these tourist attractions are intended to generate income and employment.

The most important source of income funneled into benefit-sharing comes from hunting concessions, allocated to the KA by the MET in 2006, and the first hunting concessions ceded within a national park in Namibia. With the first concessions running from 2006 to 2007, the KA 'encouraged hunters in the tender process to offer more than meat and money' and contribute to the community in more long-term ways by providing training and jobs (Koot, 2013: 134). Although a later round of concessions was more problematic (ibid.), outstanding issues were resolved and the concessions were put out for tender again for the period 2011–2015. For these two periods, 50 percent of the proceeds from the hunting concessions were held by the KA and 50 percent were paid to the Namibian Game Products Trust Fund. The proceeds held by the KA have been used to cover operating costs including salaries, to make individual payments to park residents, and to fund community projects (ibid.: 134; Jones and Dieckmann, 2014). Money transferred to the Game Products Trust Fund can also be accessed by communities in some cases, since it is used to fund conservation projects. Game meat is also distributed among the residents of the park (IRDNC, cited in Jones and Dieckmann, 2014: 391).

The KA also manages income from the harvesting of Devil's Claw. The amounts of the plant to be harvested each year are agreed by the MET in consultation with the KA to ensure sustainability. At the time of the research in 2015, the KA had negotiated a single contract for the purchase of all Devil's Claw harvested within the park in an effort to protect harvesters to some degree from price variations and guarantee a source of regular income. The KA collects the plants gathered by harvesters and stores it before sale to protect harvests from damage. Devil's Claw harvested within the park is also organically certified, which increases its value. In this sense, the KA administers a functioning benefit-sharing agreement for resident communities in Bwabwata National Park, including the Khwe.

Bushbuckridge, South Africa

Observing meetings of the Kukula Traditional Health Practitioners Association feels familiar as a scholar of social movement organizations and NGOs. There are discussions, occasional and generally polite disagreements, some evidence of frustration and burnout and a need to take breaks and take part in team-building and other activities to keep energy levels high over two days of meetings. It is obvious that this group has a clear general idea about what needs to be achieved, and determination to carry on with the hard work that has led them this far. Each meeting session opens with prayer, songs, and dance. An unexpected visit from two giraffes, grazing just outside the meeting venue, is a real surprise to all on one occasion – though these animals live close by, many Association members have never had the opportunity to see them. Protected areas are expensive and far off.

The work done by the Kukula Traditional Health Practitioners Association (Kukula) on their community protocol is already well documented (Sibuye *et al.*, 2012; Cocchiaro *et al.*, 2014; Delgado, 2016). The research focusing on the Kukula group outlined here began in 2015, some years after the group published their community protocol. At that point, the protocol had been in place since 2009, with the aim to govern external actors' use of traditional knowledge of medicinal plants held by Kukula members, and to form the basis of a 'traditional knowledge commons' based on common procedures of self-governance (Cocchiaro *et al.*, 2014). At the time of the visit, the Kukula group was considering how to move forward as an organization, and discussing both the possibility of updating their community protocol in light of increasing challenges in accessing medicinal plants on communal land, and how to move forward with specific projects aimed at securing access to plants in protected areas. The new draft of the protocol, as well as progress on these access projects and other work with protected area managers, formed the subject of discussions during a second visit in June/July 2017.

While legal backdrops are key in understanding the other cases presented so far, an important element in interpreting the Kukula's collective action arguably has more to do with overlapping layers of authority and ambiguity about which authority is responsible for what in the South African context. The layers of governance of clearest relevance here include traditional, municipal, provincial, and national authorities, as well as protected area managers. The specific challenges of the global rhino poaching crisis faced by all of these actors as well as foreign and international actors also contribute to the picture. The following paragraphs give a brief overview of these layers of governance and present the ways in which they affect the goals the Kukula group were seeking to achieve.

In South Africa, traditional leadership has a complex recent history. Findlay and Twine note that traditional leadership was exploited for local governance purposes under the apartheid regime with little local accountability, leading to lasting mistrust (2018). In democratic South Africa the role of traditional authorities has continued, particularly in rural areas: Hauck and her colleagues (Hauck, Mbatha and Wynberg, 2014: 30) explain that the Communal Land Rights Act of 2004 and the National House of Traditional Leaders Act of 2009 recognize the latter as sources of authority in rural areas, but also note that ambiguity exists about their roles and powers, and that decentralization efforts have also transferred similar powers to local government. Traditional authorities in South Africa are tasked with local support and administration in line with local customs and traditions. They take part in local development planning, may designate uses for communal land, grant communal land for different purposes, and hold joint service delivery agreements with a municipality. Traditional leaders are recognized in South Africa's Constitution, and must establish a traditional council – 40 percent of which must be elected (and 30 percent of which must be made up of women). Traditional leaders sit on municipal councils in order to allow joint decision-making, but may hold only up to 20 percent of seats in these.

During an interview in Bushbuckridge, the spokesperson of one local traditional authority described its role in terms of land use as follows. Each village assigned to a traditional authority has a community development forum and an *nduna* (a leader, see Findlay and Twine, 2018). Development forums may make proposals, and are consulted by the traditional authority if any proposal is made regarding their village. The *nduna* addresses smaller local questions that arise in villages, and liaises with the traditional authority. The council of the traditional authority considers applications for different uses of communal land within its jurisdiction – for farming, grazing, or for commercial use. Applications must contain relevant documents including maps, surveys, evaluations, and in larger cases environmental impact assessments. The council then makes a decision, in consultation with the *ndunas* and community development forums where needed, and its recommendation is then passed to municipal authorities. This is generally the end of the process unless there is a disagreement with the municipal authority. How disagreements were resolved, whether they existed between *ndunas* and the traditional council, between the council or *ndunas* and community development forums, or between these actors and municipal authorities, was unclear. Interviewees at the provincial level also underlined that the position of *nduna* was often contested given its hereditary nature, injecting more uncertainty into the institution.

In addition to this complex local network of authorities, national and provincial levels of governance also affect the interests of the Kukula Association in relation to national parks and protected areas. South African law incorporates specific provisions on benefit-sharing: the Constitution includes the right to the environment and environmental conservation as well as ecologically sustainable development and the use of nature for socially and economically justifiable development. This is given locally relevant shape in the National Environmental Management: Protected Areas Act (Act 57 of 2003), which stipulates local community involvement in the management of protected areas, and specifically mentions benefit-sharing agreements as a possible component of co-management agreements. Co-management agreements concern benefit-sharing both in a narrower sense, in that they encompass rules on how benefits produced in co-managed areas should be shared, and in a wider sense, in that co-management itself may be considered as a form of benefit-sharing. Benefit-sharing is also included in legislation on forestry, including via participatory forest management, as well as access and benefit-sharing legislation.

Access to protected areas to harvest medicinal plants is of central interest to the Kukula, who underline that plants have become scarce on communal land. Representatives of different protected areas were present at the Association meetings attended in 2015 and 2017. The protected area managers present were well informed and committed to benefit-sharing, but were also constrained in their range of action, in particular by the ongoing increase in rhino poaching in the area. Rhino poaching had become particularly problematic within the nearby Kruger National Park, which meant that park authorities had all but stopped granting local community access to protected areas – both to stop poachers gaining entry and to protect those entering from danger. There was also some discussion about whether the Kukula needed to set up internal disciplinary procedures for practitioners indirectly supporting poachers, for example by giving treatments intended to enhance potency. Although there was no suggestion that Kukula members had done so, the Association members agreed that internal procedures would be a good way of preventing this indirect support, as well as communicating the importance to members of passing information about poachers to park managers or other authorities.[14] In addition to the poaching crisis, protected area managers highlighted that access to plants was also affected by the complex legal framework applying to protected areas, which includes plant health and protected species legislation. Several solutions suggested by various protected area managers to the question of access to medicinal plants thus foresaw access to plants without practitioners actually entering protected areas.

Another potential source of complexity in terms of access to protected areas also flows from the national and provincial levels. Following the end of apartheid, the South African government provided for groups to reclaim lands that had been taken from them through the Restitution of Land Rights Act (Act 22 of 1994). The window for filing claims remained open until the end of 1998. However, because of a large backlog, the window was reopened in 2014 for another five years. Wherever lands subject to claims fall within (or partly overlap with) protected areas, specific rules apply. Land within national parks cannot be restored, and claimants are given compensation and special access rights instead. Where land falls within the boundary of a provincial park, claimants usually enter into a co-management agreement with the provincial authority, including benefit-sharing measures. Claimants can also take over management entirely, though the most common approach is co-management, with progressive transfers of authority to the claimants over time, according to representatives of provincial authorities. Rules on what activities are allowed in protected areas remain in place where land is reclaimed, which means that protected areas continue to be subject to complex rules. One of the protected areas that the Kukula were working to gain access to in 2015 was subject to a land claim, adding a further layer of complexity due to potential conflicting interests. The situation is further complicated by the fact that the Bushbuckridge area falls within the Kruger to Canyons UNESCO biosphere reserve (K2C).

The Kukula Traditional Health Practitioners Association

The Kukula members' aims to access medicinal plants are based on their role as traditional health practitioners. Some background on the Association and the role its members play in their communities is useful before looking at these aims in more detail. The traditional health practitioners that form the membership of the Kukula Association are all from the wider Bushbuckridge area, which stretches across two provinces, Mpumalanga and Limpopo, close to the Kruger National Park. People living in the area come from a number of different groups with different traditions and languages: Tsonga/Shangaan, Swazi, Tswana, Bapedi (North Sotho), and Basotho (South Sotho). Unemployment and poverty are high in the area, as are the numbers of those living with HIV (Cocchiaro *et al.*, 2014). In 2009 a group of traditional health practitioners from this area worked together to establish the Vukuzenzele medicinal plant nursery. They also began to collaborate with Natural Justice and the K2C management committee, hoping for their assistance in founding an association. They had undertaken the work on the plant nursery with a view to securing access to medicinal plants, and an association would be of help in this vein, as well as for securing ministerial recognition for their work, thus allowing them to receive and refer patients. In fact, palliative care was seen as a particularly important role for traditional healers, particularly given their practice of not asking to be paid. An association was also seen as a good basis for drafting a community protocol to protect traditional knowledge and outline the need for and role of traditional practitioners in protecting threatened medicinal plant species.

The community protocol was duly drafted by the Association's members with support from Natural Justice and the K2C management committee as well as a local law firm (Traditional Health Practitioners of Bushbuckridge, 2010). The Association also took part in various projects with and was supported by academics at the University of the Witwatersrand Rural Facility in the Bushbuckridge area. The protocol provides details of the traditional practitioners' contribution to their communities, principally their contribution to the health of communities (ibid.). On this point, the practitioners acknowledge that their approach to medicine is not universally accepted. However, they emphasize that they do not ask for any payment for their services, and provide not only care but counseling to local communities. Traditional practitioners also refer patients to clinics where possible, hence their wish to gain official recognition from the Ministry of Health. Traditional practitioners are also important holders of traditional knowledge with different specializations. The group used their community protocol to protect this knowledge by creating a commons, pooling their traditional knowledge in order to better protect it. Not all members share their knowledge with others. Rather, they have pooled their knowledge to regulate access according to common rules, including those on training new practitioners. Other rules concern sustainable harvesting techniques – the traditions of the practitioners govern the methods used to gather plants. In particular, the strength of any medicinal preparation is linked to whether the plant the material came from is still living. Sustainable use is thus inextricably linked to traditional knowledge. Finally, the traditional practitioners play an important cultural role. They are seen as having unique connections with ancestors, and are key in naming ceremonies and other rites of passage.

Over time, the Association's role and activities have evolved. In 2011, the Association signed a non-disclosure agreement with a local cosmetics and bedding company, agreeing to supply plants with potential commercial value. Although the agreement was ultimately unsuccessful, this event signaled the growing efficiency of the group. From 2013 to 2015, the Association's members took part in a Global Environment Fund project alongside staff from the University of the Witwatersrand Rural Facility and the K2C management committee. This project aimed to develop an evidence-based protocol for the sustainable harvesting of medicinal plants that led to the identification of the Mariepskop Forest Reserve as a useful potential source of plants. The Association also held meetings with officials from the Ministry of Health and began to discuss how to proceed with the development of an internal code of conduct. By 2015 when the first research visit took place, the Association had grown from its original 80 members to count more than 300. At the meetings observed in 2015, the Kukula members present discussed their ongoing approach to challenges linked to access to medicinal plants.

These challenges included the overharvesting of medicinal plants on communal land (a theme already introduced in their community protocol). This is seen to result from unregulated access and the unsustainable harvesting practices of suppliers to large urban markets in Durban and Johannesburg. Connected to this, traditional health practitioners also felt excluded from decisions by local traditional authorities over questions of land use designation. They felt that too much local land had been given over to grazing, agriculture and commercial uses, exacerbating medicinal plant shortages. Finally, the practitioners felt excluded from protected areas – including national and provincial parks, forest reserves, and private reserves. They understood that plants were more plentiful in these locations, but saw barriers ranging from physical fences to complex rules and regulations as excluding them. A key aim for the Kukula in 2015 was to gain recognition from a variety of actors. This was seen as both a first necessary step towards understanding the complex rules that needed to be navigated to achieve their aims, and a good way of showing their commitment as a serious professional Association. At meetings in 2015 the Association thus began a dialogue with a traditional authority and protected area managers from both the Kruger National Park and Mariepskop Forest Reserve.

These meetings allowed different actors to better understand the position of the Association's members, and for the Kukula to reach a better understanding of the constrictions in place on others. These conversations led, as mentioned, to suggestions about how to solve plant access issues without entering protected areas. By cultivating medicinal plants outside protected areas, the Kukula would both benefit and contribute to re-establish biodiversity. Nevertheless, it should be noted that not all of the plants of interest to the Kukula can be cultivated outside protected areas. Many issues that could hinder these solutions were also raised. For example, protected area managers required lists of the species practitioners wished to access before any visit to the reserve could be arranged. The practitioners explained the difficulties of providing such lists: some were related to the way in which they practice. Certain plants – sometimes from specific locations – come to practitioners via their ancestors in dreams. Many practitioners only harvest enough of a plant to treat a particular patient at any one time, making it challenging to predict their needs for particular plants. Language also proved challenging, since practitioners did not always know the scientific names of the plants required. Arrangements for protected area managers and Kukula members to go for a veld walk to identify and name plants together were thus discussed. Understanding these difficulties, in addition to practical questions about how and what to share among their 300 members, how to store and process plants and the like, also contributed to the alternative suggestions made at the meeting.

Moving forward, in 2017 the Kukula Association had continued its active and varied approach to developing new relationships. Work with the K2C management committee had continued and this organization underlined its support for the Association. Members of the Association had also taken part in training courses run by SANParks on vultures, aimed at tackling problems linked to the poisoning of these birds by poachers (their presence can alert park rangers to a carcass) and consequent health risks linked to their subsequent use by some practitioners. Another SANParks project supplying Pepper Bark trees to be grown outside protected areas also saw key involvement from Kukula practitioners. The manager of the Vukuzenzele medicinal plant nursery played a central role in distributing and training others about how to care for these trees. Association members had also taken part in consultations about Kruger National Park's next ten-year management plan, in line with the park's rules about community consultation and benefit-sharing. Leading members of the Association had also attended a South African National Biodiversity Institute (SANBI) workshop in the neighboring city of Nelspruit to discuss endangered plant species in the area, and the possibility of involving them as members of the Steering Committee was discussed.

At the Association meetings observed in 2017, one of the main items of business was to discuss a possible land purchase and how to use the land. Ideas included a plant nursery, or the foundation of a school and treatment center. The latter was suggested as a method to boost youth involvement in the Association. Another item discussed was the application for access to Maricpskop Forest Reserve, which had been submitted but met with various hurdles at different governance levels. Association members also heard and debated presentations from various actors, including a presentation by the author on the preliminary findings of the research project this book is based on. A clear issue arising from these presentations was that the Association's members felt they had done much to engage with a variety of local actors and international researchers to build mutually beneficial relationships, yet rarely heard back from them. More than one participant commented that the international researchers present at the discussions were the first to return and report back to the group on their findings and how they might relate to the Association's aims and activities. Finally, the members discussed the revision of their community protocol, a theme that had been taken up at various meetings since 2015. The revised protocol was published in July 2018, adding an Annex to the original text outlining legislation that supports the Kukula's position drafted in consultation with a local law firm. The updated protocol also reiterates calls for dialogue between the Association and a variety of local actors.[15]

Bario, Malaysia

Traveling to Bario by air, a clear change in the landscape is visible where palm oil plantations end and forests begin, lined with ever fewer logging roads. The 12-seater plane lands in Bario airport, where some of the staff are also rice farmers. The small airport restaurant is recommended – the pilots always eat their lunch here because of the town's reputation for good food. Visiting the original Bario longhouse, photographs on the walls bring home the extent of change in a single generation, as younger family members in suits and dresses pose alongside parents in traditional Kelabit costume. As the food festival begins, the previously quiet town fills with people. Dignitaries flown in by helicopter are welcomed by schoolchildren holding the famous Bario pineapples and deliver long speeches. They are treated, along with all the other visitors, to displays of traditional dance, music, and games, Bario rice with vegetables and meat from the surrounding forest. Many Kelabit return to Bario for the food festival, and there is a sense of celebration and pride in the many achievements of the town overlooked by Prayer Mountain.

Bario, in the highlands of the Malaysian state of Sarawak, Borneo, is home to a unique variety of rice. Grown by the local Kelabit community, the rice is prized for its quality internationally, and plays a central role locally in both food security and culture.[16] As more Kelabit have left the Bario area in recent generations, cultivation of the rice has fallen due to a combination of this local out-migration and a fall in Indonesian immigration. To address this, a tripartite agreement between the local community association (Rurum Kelabit Sarawak – RKS), the Ministry of Agriculture of the Sarawak sub-State government, and an agricultural company was concluded. The agreement was intended to provide farming services to those who wanted them, as well as to improve farm roads and irrigation. Two visits were made to Bario during the annual local food festival held from late July to early August in 2015 and 2017. The food festival sees many local community members that live elsewhere return home, while politicians and other stakeholders are also invited, providing the opportunity for the community to connect with a range of relevant actors. Before looking at the tripartite agreement in more detail and considering a range of issues raised amongst community members and others, some more general information about the governance context and the history of the Bario area and the Kelabit is useful. As in other cases outlined here, history and governance are closely bound up together and provide a key to an understanding of the case.

Bario is, as mentioned, located in the central highlands of Borneo, close to the border with the Indonesian State of Kalimantan. The Federation of Malaya, which united the peninsular states of current-day Malaysia, was formed in 1957. The Borneo States of Sarawak and Sabah (as well as Singapore, which later left) joined the Federation in 1963 to form the current Federal State of Malaysia (a constitutional monarchy). Sarawak has a unique history and identity within Malaysia. Its entry into the Federation of Malaya included a Sarawak constitution that apportioned more powers to its State government compared to others in peninsular Malaysia. This was intended to protect its population, unique in Malaysia in terms of its ethnic and religious composition. Over time, the State was progressively integrated into the country's more general government system, though it retains its uniqueness, particularly in terms of the mix of faiths (such as the Kelabit, who are mostly Christian) and the widespread use of English (Porritt, 2007).

Bario is in turn unique within Sarawak. It is the central village of what are known as the Kelabit Highlands. The Kelabit remained rather separate from the rest of the State's population until relatively recently, given the difficulties of accessing the interior of Borneo. During the Second World War, Bario was used as a base by allied troops (including Major Tom Harrisson, who remained linked to the area) in the fight against Japanese forces, and Kelabit people played an active part. An airstrip was built at this time and was central in opening Bario to the rest of the country, ushering in an era of swift change. In 1963 the local population was boosted when the conflict with Indonesia saw Kelabit people living close to the border relocated. This combination of an increased population and accessibility cemented Bario as the center for administrative, health, and education services in the Kelabit Highlands (Janowski, 1991). Nevertheless, traveling to Bario by land has only become possible in recent years via logging roads, and still only with 4x4 vehicles. Some local roads to outlying villages in the area are also of very recent construction.

The Kelabit have a customary system of local governance and decision-making. Each village (the Bario area covers a total of 13 villages, each organized around a historic longhouse) has a head, and the heads of each village form the Council of Elders. The Pemanca (or Penghulu) leads the Council of Elders and is the head of the Kelabit people. In more recent years, this customary system of local governance has weakened as many of the more experienced Council members have passed away. Their successors, where one is named, have often lived elsewhere for many years for reasons of study, work, or both, and some community members recounted they felt they lacked the knowledge, training, and experience required to fill the role. At the time of the research, the RKS had been asked by the Council of Elders to liaise with the Sarawak State government about the tripartite deal for rice agriculture. The group was also running the Bario food festival – an important annual event demonstrating not only local culture and food, but also the use of the surrounding lands – an important element in keeping logging companies at bay. The RKS had thus assumed some aspects of the role once fulfilled by customary governance alone, though only when requested to do so by the Council of Elders. More information about the RKS is thus useful. Registered in 1994 as an NGO active on behalf of the Bario community on economic and social issues, sports, education, culture and the arts, individual chapters of the RKS exist in a number of cities in Sarawak.[17] The group is led by a State Council, with a president elected every three years at meetings attended by the heads of each chapter. Although forming a NGO to liaise with the State government on behalf of a particular community is common in Malaysia, given that there is no municipal government level, the RKS also reflects Kelabit community culture. Janowski (1991) describes the community's strong sense of attachment and duty to Bario, and how Kelabit who live and work elsewhere dedicate time and other resources to work for its improvement.

Other features of the local community also speak to a broader understanding of the tripartite agricultural deal. First, in connection with the high proportion of Kelabit community members that leave the Bario area to pursue their studies and careers, and in some cases go on to make up the ranks of the RKS, it is worth noting that a comparatively high proportion of these community members work in high-level positions in Sarawak and beyond. Many hold postgraduate university degrees, for example, and several hold positions in the Sarawak government. The academic success of the local population is a matter of pride in the Bario area, and attributed not only to work ethic but to the healthy way of life of the community and, in particular, the properties of local Bario salt (see Zaleha *et al.*, 2002). The positions held by various RKS members were seen as one element in the community's success in securing funding from the Sarawak State government for the tripartite agricultural agreement. The number of people moving away from the area is also, however, a factor in the decline in rice farming – the local population is an aging one. Another point to note is the non-confrontational culture that characterizes Kelabit society, despite the distinctive identities attributed to each of the longhouses that historically formed the different villages in the area. Issues arising in the local community have traditionally been solved via consensus and innovation.

A final point of relevance is an early draft of a Kelabit community protocol. The idea of drafting a community protocol has been under consideration for a number of years within the community, and a draft was produced following a workshop held at the 2013 e-Bario knowledge fair.[18] These more or less annual fairs involve both researchers from the University of Malaysia Sarawak and the local Bario community, as well as invited speakers, and the 2013 event included a workshop on community protocols led by a representative of Natural Justice. A draft protocol followed, providing information on Bario's geography, on Kelabit culture and language, and guidelines for visitors to the area. The focus of the document at present is on tourists and non-commercial researchers. Although it was not under general discussion within the community at the time of the research, its focus does provide more background to the issues raised around the tripartite agricultural agreement including traditional knowledge and practices in the cultivation of Bario rice.

Bario's tripartite agricultural deal

In Bario it was the local community that sought an intervention from government and private actors to help reverse the decline in the cultivation of Bario rice. The tripartite agricultural deal was not the first attempt to achieve this goal – the Bario Rice Certification Scheme introduced some years prior had also tried to revitalize production. Aimed at promoting on-farm conservation, a 2007 country report prepared for the UN Food and Agriculture Organization mentions the scheme and describes local rice production as traditional, with no use of pesticides or chemicals.[19] A certificate of geographical indication awarded to Bario rice in March 2009 also confirms this. The certificate is awarded 'where a given quality, reputation or other characteristic of the goods is essentially attributable to their geographical origin'.[20] In addition, Bario rice received a Presidia Award from the Slow Food Foundation, whose involvement included a project to support traditional cultivation methods that ended in 2011.[21] Neither of these efforts resolved the issue of falling rice cultivation levels, but both underlined an understanding of Bario rice as a traditional and organic product. Indeed, according to traditional methods, local Bario rice growers plant and harvest by hand, and use no pesticides or chemical fertilizers on the rice crops themselves (though some are used on the edges of fields).

The decision to attempt a different type of project thus followed other efforts. The project was pursued by the RKS as it made a good fit with the Malaysian government's 2010 Transformation Program,[22] which included National Key Economic Area projects. The Bario Rice Industry Development project was secured within this framework as an Entry Point Project, and RM17 million was allocated in government funding.[23] An 'anchor farmer', a farming company able to deliver the agricultural services foreseen in the project, was then selected. The company chosen was Ceria due to its similar work elsewhere. Ceria formed a subsidiary company to deliver the project, Bario Ceria Sdn Bhd. Ceria owns the company with a 51 percent share, and the RKS has a 49 percent share. The aim of the project was to provide a range of services and infrastructure to improve rice production. Mechanized leveling, plowing, planting, and harvesting services would be made available for local farmers, or landowners could opt for Bario Ceria to manage cultivation entirely. Machinery was effectively owned by the government and, according to interviewees, will be transferred to the community upon completion of the project. To support these services, the project also included infrastructure improvements, envisaging the construction of seven irrigation dams to complement existing systems, assistance for their upkeep and repair,[24] and the construction of farm roads and a drying and milling factory.[25]

The project began in 2013 for an initial area of 200 hectares of agricultural land around Bario. An extension of two years and 200 hectares was confirmed in 2015, and in 2017 a further extension for as long as ten years was rumored to be under discussion. The farm roads and irrigation dams had all been completed in 2015, and the first harvest of rice from those fields covered by the agreement was available for sale in early 2014. Production- and profit-sharing was fixed at 70 percent/30 percent for land managed entirely by the farming company. Landowners received 30 percent of the rice harvest from their land, while 30 percent of any eventual profits from sales were transferred to the RKS for community use.

Two main issues can be linked to benefit-sharing in the Bario case: profit- and production-sharing on one hand, and farming techniques on the other.

Issues raised with regard to both of these are linked to the theme of consultation. Before discussing the issues raised by some community members, however, it is worth underlining that the benefits gained and shared from the project across the community as a whole were widely recognized, and that views on the deal have evolved since the beginning of the project. Some community members who were initially skeptical came to support the project, while others who were initially enthusiastic later reconsidered.

As noted above, profit- and production-sharing ratios were fixed at 70/30 following a long-standing practice already applied in the past for migrant workers who cultivated rice on behalf of landowners. Since those working the land assumed most of the risk, this arrangement was deemed fair. Under the tripartite deal, however, this work is subsidized with government funds, leading some to question the arrangement. Beyond discussions about differences between local farmers' and the project farmers' sale prices, this point was raised with reference to food security issues and disappointment in the yields from the first harvests. For those landowner families who rely on their production share of rice, low yields are a real risk. This links in with issues raised about farming techniques, discussed below. Views on profit-sharing pointed instead to the theme of intra-community benefit-sharing. Profit shares from the project are transferred to the RKS, but there was little clarity about amounts transferred or about how monies would be used. No mistrust in the RKS was expressed, rather, the lack of information seemed to be linked to a surfeit of discussion forums within the project and the community. A similar lack of clarity characterized views expressed about the long-term effects of the project and expectations about training in the use of farm machinery and jobs for local youth. It should however be noted that the aim of the project as a whole was to boost rice cultivation rather than generate monetary profit.

The second major area of discussion within the community concerned the techniques employed by the farming company. As noted, Bario rice growers use traditional methods, while the farming company took a different approach. In particular, it experimented with more intensive production techniques, used industrial fertilizers and pesticides, and, according to some, grew varieties other than Bario rice. Each of these points can be discussed in some more depth.

At the beginning of the agricultural management project, the farming company experimented with intensive production techniques that differed from traditional methods based on specific local knowledge. The traditional methods used by local Kelabit farmers are geared to producing one rice harvest per year in line with their knowledge of the local climate, pests, and other risks. Traditional farming approaches also include leaving water in the fields throughout the season, allowing the remains of rice plants to rot and thus keep the ground fertile. Water buffalo are also used to improve soil fertility. In addition, keeping the fields flooded allows farmers to keep fish in the water – another food source. The farming company's approach at the beginning of the project was different. Rather than keeping fields flooded, they were drained and industrial fertilizer used before planting the rice. Topsoil was then scraped away after the harvest and the process repeated in an attempt to achieve two rice harvests per year. The reasons for these differences are likely linked to the key performance indicators attached to the government subsidies and the consequent need to increase yields over time.

Whatever the reasons behind them, these more intensive farming techniques provoked concern in sections of the community. Some landowners and farmers withdrew from the scheme, as they felt that these techniques were not appropriate for the Bario area. The relatively poor yields in the first years of the project seem to support this view, as does the fact that the farming company later changed its approach to a single harvest per year. There is a clear link to consultation themes here, as well as the perception of the worth and relevance of traditional knowledge.

The use of industrial fertilizers and pesticides also raised some concerns in Bario. While traditional farming techniques mostly rely on other methods – water buffalo and leaving harvested plants to rot – the farming company instead used both industrial fertilizers for the soil and industrial pesticides on crops. These practices were a cause for concern not only with regards to soil fertility, but also the spread of these products to adjacent, traditionally farmed fields and a subsequent decline in fertility in these areas too. Some community members felt that pest numbers had risen in traditionally farmed fields, and linked this to pesticide use. Combined with the planting of other rice varieties, the use of industrial fertilizers and pesticides also raised some questions about the continued use of the certificate of geographical indication. This was not, however, mentioned by local community members themselves. Regarding the farming company's use of other rice varieties, concerns centered on the fact that one of the original reasons behind the deal was to conserve the Bario rice variety. Though the farming company's initial approach involved planting other rice varieties, a Bario rice seed nursery was later established. This choice may have been linked to supply problems given the context of falling levels of rice cultivation: there may simply not have been enough seedlings available to supply the project at first. Nevertheless, the presence of different varieties could lead to cross-pollination, with knock-on effects for the Bario rice variety, and the reasons for using other rice varieties were in any case not clearly communicated. Overall, many of the concerns about farming practices were addressed by the farming company as the project progressed. The planting calendar was confirmed to be in line with traditional practices at the time of the first research visit, and seed nurseries had been established. There was no indication that the use of industrial pesticides and fertilizers would be reduced however.

An important link between the issues raised is that they all relate to questions of discussion and consultation. Many of the issues were resolved precisely through dialogue between the farming company and local community members, suggesting that some might have been avoided through earlier consultation at the project design stage. This could conceivably apply, for example, to production- and profit-sharing ratios, as well as the use of profits. Instead, community members told a story where there was widespread support for the tripartite agricultural management project and trust in the RKS, and local community members were kept informed in a general rather than a detailed way as negotiations progressed. The reasons behind this approach may include issues about the distance between Bario and the seat of the Sarawak State government, and, perhaps more convincingly, the fact that the RKS was not created with a structure designed for this type of role, leaving a gap for a forum for consultation with local farmers.

An interesting point emerged in this vein regarding the lengths of time it took for the farming company to make changes and communicate these. A number of forums where community members could raise their concerns existed in Bario, and this appears to have created some confusion and slowed down communication in both directions as well as the implementation of solutions. Fields managed by the farming company are grouped into blocks, and each of these is assigned a block leader. Some community members raised concerns with block leaders, who then liaised with the farming company. Other community members raised issues directly with the farming company, both locally and on occasion in the city of Kuching, where the company has an office. Still others raised concerns with the RKS, who in turn liaised with the company. In 2017, a WhatsApp group of local community residents and RKS members (with the two groups partly overlapping) was mentioned as another such a forum. Thus, procedures for raising questions or concerns linked to the agriculture project were unclear, not because of a lack of spaces for exchanging opinions, but because of a surfeit. This idea is bolstered by the fact that some community members confirmed that changes had been made, while others did not, once again suggesting a problem in communication. As the project progressed, efforts were made in this direction between block leaders and the RKS, since the former underlined the need for frank conversations about where to address issues raised about the agricultural project. There was also some discussion about how farmers themselves could organize, for example by forming and registering an official farmers' association.

Different stories, common themes

Each of these communities' stories is unique and complex. These brief case studies do not claim to capture these intricacies, nor do they seek to suggest that the realities of these situations can be grasped through such short research visits. Rather, the aim of this chapter was to convey some of the main features of discussions related to benefit-sharing across a range of different communities and situations. Any broad themes that emerge from

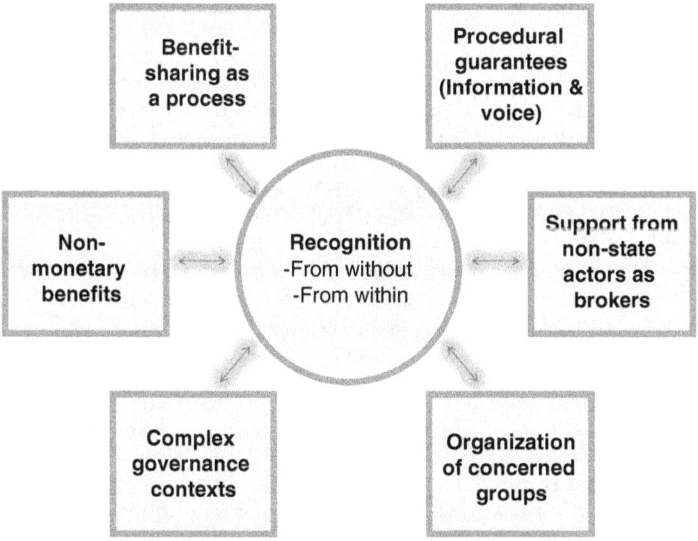

Figure 2.1 Common themes arising from the case studies.

the cases could thus be considered fruitful avenues for an analysis of the international level of environmental governance.

A number of common themes do in fact emerge from a comparison of these cases. The next chapter takes a fresh look at the case studies, delving into them to illustrate how each of the common themes detailed in Figure 2.1 comes to the fore.

Notes

1 See e.g. Marissa Tefada, 'The Greek island with the key to longevity', BBC. www.bbc.com/travel/story/20171116-the-greek-island-with-the-key-to-longevity (accessed 1 June 2018).
2 National Constitution of Argentina, art. 75, point 17: 'To recognize the ethnic and cultural pre-existence of indigenous peoples of Argentina.'
3 National Constitution of Argentina, art. 41: 'All inhabitants enjoy the right to a healthful, balanced environment fit for human development, so that productive activities satisfy current needs without compromising those of future generations, and have the duty to preserve the environment...'
4 The communities brought a writ of amparo before the Argentine Supreme Court in December 2012. They also brought their complaint to the attention of the UN Special Rapporteur on indigenous rights at the time, James Anaya, who visited the area in late 2011, and to the UN Committee on Economic, Social and Cultural Rights in the same year. In June 2013 their complaint was brought before the Inter-American Commission of Human Rights.
5 Specifically, the proposed routing of the Dakar Rally through their lands for the 2016 race.
6 Personal communication with representative of FARN, June 2019.

7 I refer to the 'Khwe' in line with Jones and Dieckmann (2014) as the preferred term among this group. Others refer to the group as the 'Khwe San' or 'San Bushmen'.
8 Information on the history of the park is drawn from Jones and Dieckmann (2014).
9 The name Caprivi is that of the German Chancellor in office in 1890 when the Heligoland-Zanzibar Treaty was signed: Leo Von Caprivi. This treaty between the UK and Germany integrated this strip of land, allowing access to the Zambezi River, into colonial 'German South West Africa'.
10 !Xun San and Mbukushu communities also live in the area, as discussed later (but see Koot, 2013; Jones and Dieckmann, 2014).
11 A factor here is that the Khwe usually have no single traditional leader, given their traditionally horizontal organization. See Vermeylen (2007).
12 See NACSO (2015).
13 See Natural Justice and the ABS Initiative (no date).
14 However, it is worth stressing that the risks associated with reporting information about poachers were and remain significant, raising questions about how a local association could be reasonably expected to contribute to this issue.
15 See https://naturaljustice.org/the-kukula-traditional-health-practitioners-association-release-their-updated-biocultural-community-protocol/ (accessed 1 April 2019). The revised protocol is available at https://naturaljustice.org/wp-content/uploads/2018/06/BCP-Kukula-Traditional-Health-Practitioners-Association-English.pdf (accessed 1 April 2019).
16 Janowski (1991) describes rice as central to important Kelabit naming ceremonies that accompany parenthood and grandparenthood. Rice is also considered the staple foodstuff in the area. Its economic importance also grew from the 1960s on.
17 http://rurum.org.my (accessed 17 November 2015).
18 The e-Bario knowledge fair project provided computers and ICT services to the Bario area, including free Internet in the center of Bario and a radio station. See www.ebario.org (accessed 29 November 2015). It has since been expanded to include other indigenous groups and renamed the e-Borneo knowledge fair. See ebkf.org (accessed 26 April 2019).
19 Field Document No. 2007/01 prepared for UN FAO, available at www.fao.org/docrep/013/i1500e/malaysia.pdf (accessed 2 October 2015).
20 According to the Intellectual Property Organization of Malaysia, at www.myipo.gov.my/petunjuk-geografi (accessed 20 November 2015).
21 On these projects see: www.fondazioneslowfood.com/wp-content/uploads/2015/06/ING-bilancio-sociale20071.pdf.
22 See www.pemandu.gov.my/gtp/About_GTP-@-GTP-;_The_Story_So_Far.aspx.
23 www.theborneopost.com/2014/01/16/nkea-first-harvest-of-bario-rice-on-sale-soon/.
24 www.theborneopost.com/2013/10/06/bario-rice-enjoying-new-lease-of-life/ This article also sees Bario Ceria acknowledge that local farmers remain to be convinced by the project and that issues have been raised.
25 www.theborneopost.com/2014/01/16/nkea-first-harvest-of-bario-rice-on-sale-soon/ (accessed 17 November 2015).

References

Anaya, J. (2012) *Report of the Special Rapporteur on the rights of indigenous peoples, Progress report on extractive industries*. UN Doc. A/HRC/21/47.

Bareli-Gaglia, M. (2010) 'The Ikarian paniyiri, the commons and the gift', in *72nd Annual Meeting of the Society for Applied Anthropology – Bays, Boundaries and Borders.* Baltimore.

Cocchiaro, G. *et al.* (2014) 'Consideration of a Legal Trust Model for the Kukula Healers' TK Commons in South Africa', in de Beer, J. *et al.* (eds) *Innovation and Intellectual Property: Collaborative Dynamics in Africa.* Cape Town: University of Cape Town Press, pp. 151–170.

Comunidades Indígenas de las Salinas Grandes y Laguna de Guayatayoc de Jujuy y Salta (2015) *Kachi Yupi/Huellas de la Sal: Procedimiento de Consulta y Consentimiento Previo, Libre e Informado para las Comunidadades Indígenas de la Cuenca de Salinas Grandes y Laguna de Guayatayoc.* Available at: https://cl.boell.org/sites/default/files/protocolo_final.pdf.

Delgado, N. A. (2016) 'Community protocols as tools for resisting exclusion in global environmental governance', *Revista de Administração de Empresas*, 56(4), pp. 395–410.

Dinan, D. (2010) *Ever Closer Union. An Introduction to European Integration.* Fourth edition. Basingstoke: Palgrave Macmillan.

Findlay, S. and Twine, W. (2018) 'Chiefs in a democracy: a case study of the "new" systems of regulating firewood harvesting in post-apartheid South Africa', *Land*, 7(1), p. 35. doi: 10.3390/land7010035.

Hauck, M., Mbatha, P. and Wynberg, R. (2014) 'Coastal Communities and Livelihoods in South Africa and Mozambique', in Wynberg, R. and Hauck, M. (eds) *Sharing Benefits from the Coast: Rights, Resources and Livelihoods.* Cape Town: University of Cape Town Press, pp. 17–36.

Gaglia Bareli, M., Geelhoed, M., Parks, L., Morgera, E. and Tsioumani, E. (forthcoming 2020) 'The unintended consequences of the Common Agricultural Policy for local communities', in Cardwell, P. and Grangers, M-P. (eds) *Research Handbook on the Politics of EU Law.* Cheltenham: Edward Elgar.

International Labour Organization (1989) *Convention no. 169 Concerning Indigenous and Tribal Peoples in Independent Countries.* 28 ILM 1382.

Janowski, M. (1991) *Rice, Work and Community among the Kelabit of Sarawak, East Malaysia.* London School of Economics.

Jones, B. (2010) 'The Evolution of Namibia's Communal Conservancies', in Nelson, F. (ed.) *Community Rights, Conservation and Contested Land: The Politics of Natural Resource Governance in Africa.* London: Earthscan, pp. 106–120.

Jones, B. (2012) *An analysis of international law, national legislation, judgements, and institutions as they interrelate with territories and areas conserved by indigenous peoples and local communities: Report No. 4 Namibia.* Bangalore: Natural Justice. Availble at: https://www.iccaconsortium.org/wp-content/uploads/2015/08/legal-review-4-namibia-2012-en.pdf.

Jones, B. and Dieckmann, U. (2014) 'Bwabwata National Park', in Dieckmann, U. *et al.* (eds) *Scraping the Pot: San in Namibia Two Decades after Independence.* Legal Assistance Centre and Desert Research Foundation of Namibia, pp. 365 398. Available at: www.lac.org.na/projects/lead/Pdf/scraping.pdf.

Jones, B. T. B., Diggle, R. W. and Thouless, C. (2015) 'From Exploitation to Ownership: Wildlife-Based Tourism and Communal Area Conservancies in Namibia', in van der Duim, R., van Wijk, J., and Lamers, M. (eds) *Institutional Arrangements for Conservation, Development and Tourism in Eastern and Southern Africa: A Dynamic Perspective.* New York: Springer, pp. 17–38.

Jones, B. T. B. and Weaver, L. C. (2009) 'CBNRM in Namibia: Growth, Trends, Lessons and Constraints', in Suich, H., Child, B., and Spenceley, A. (eds) *Evolution and Innovation in Wildlife Conservation: Parks and Game Ranches to Transfrontier Conservation Areas.* London: Earthscan, pp. 223–242.

Koot, S. P. (2013) *Dwelling in Tourism: Power and Myth amongst Bushmen in Southern Africa.* University of Leiden: African Studies Centre.

Marchegiani, P., Hoglund Hellgren, J. and Gomez, L. (2018) *Lithium extraction in Argentina: a case study on its social and environmental impacts.* Available at: https://goodelectronics.org/wp-content/uploads/sites/3/2019/05/DOC_LITHIUM_ENGLISH.pdf.

Marchegiani, P., Morgera, E. and Parks, L. (2019) *Indigenous peoples' rights to natural resources in Argentina, the challenges of impact assessment, consent and fair and equitable benefit-sharing in cases of lithium mining*, BENELEX Working Papers, 19. Available at: https://papers.ssrn.com/sol3/papers.cfm?abstract_id=3317375.

NACSO (2015) *The state of community conservation in Namibia – a review of communal conservancies, community forests and other CBNRM initiatives (2015 Annual Report).* Windhoek. Available at: www.nacso.org.na.

Natural Justice and the ABS Capacity Development Initiative (ed.) (no date) *Community Protocols in Africa: Lessons Learned for ABS implementation.* Available at: https://naturaljustice.org/wp-content/uploads/2018/11/2018_Community-Protocols-in-Africa_Lessons-Learned_Natural-Justice.pdf.

Papalas, A. J. (2005) *Rebels and Radicals. Icaria 1600–2000.* Wauconda, IL: Bolchazy-Carducci Publishers.

Porritt, V. L. (2007) 'Constitutional change in Sarawak 1963–1988: 25 years as a State within the Federation of Malaysia', *Borneo Research Bulletin*, 38, pp. 159–171.

Sibuye, R. *et al.* (2012) 'The Bushbuckridge BCP: traditional healers organise for ABS in South Africa', Special issue: Biodiversity and Culture: Exploring Community Protocols, Rights and Consent, *Participatory Learning and Action, 65.* Available at: http://pubs.iied.org/pdfs/G03403.pdf.

Traditional Health Practitioners of Bushbuckridge (2010) *Biocultural Protocol of the Traditional Health Practitioners of Bushbuckridge.* Available at: http://community-protocols.org/wp-content/uploads/documents/South_Africa-Bushbuckridge_Biocultural_Protocol.pdf.

Vermeylen, S. (2007) 'Contextualizing "fair" and "equitable": the San's reflections on the Hoodia benefit-sharing agreement', *Local Environment,* 12(4), pp. 423–436.

Zaleha, M. I. *et al.* (2002) 'Micronutrients and its correlation with mental performance among schoolchildren in Bario, Sarawak: a preliminary study', *Medical Journal of Malaysia*, 58(3), pp. 309–319.

3 Common themes in the five local experiences of benefit-sharing

This chapter explores some common themes that arise from a comparison of the cases of local discussions concerning matters linked to benefit-sharing, as outlined briefly at the end of Chapter 2 (see Figure 2.1). Because the cases are very different, the common themes that they share may, arguably, be likely to apply to other local communities. This is also suggested by the fact that there is some confirmation for these themes in existing literature on local experiences of benefit-sharing (e.g. Wynberg and Hauck, 2014; Tysiachniouk and Olimpieva, 2019). Beyond any more general claim of applicability, the common themes add to existing work that suggests how to approach benefit-sharing discussions at local levels, and indicate areas that are interesting to consider in an analysis of the international level of environmental governance. This will be discussed in depth with reference to the CBD in Chapter 4. Here, each of the common themes is illustrated with examples from the local case studies introduced in the previous chapter. The chapter thus deepens the investigations of the case studies, following a thematic approach.

The first idea to emerge from the comparison is recognition, understood in a variety of ways in relation to both external actors and within communities themselves. This is a central and more abstract theme that all the others are linked to in different ways, and is thus discussed first. Next, the chapter deals with the common themes of support from non-state actors and the organization of communities. These themes are argued to be necessary, alongside recognition, for communities to begin pursuing their goals (including eventual benefit-sharing) in a practical sense. Complex governance backdrops and procedural guarantees are then broached. All of the cases take place in complex contexts, and this complexity translates into a need for external actors to guarantee conditions for procedures if communities are to be able to act to secure their aims. Finally, the themes of non-monetary benefits and the view of benefit-sharing as a process are discussed. While the other common themes are more related to what local communities see as necessary for any eventual benefit-sharing to take place in fair and equitable ways, these last two are directly related to how local communities frame benefit-sharing.

As discussed in Chapter 1 (Comparative local case studies), the case comparison is guided by Ribot and Peluso's (2003) theory of access, which involves a focus on benefit-sharing, and sees access as 'the ability to benefit from things' (ibid.: 153). This approach involves paying attention to access and the actors and institutions with power over it. This broader view matches the exploratory approach of the case studies, which aimed to identify what local communities link to benefit-sharing beyond any existing definitions, particularly those used at the international level. The case study comparison, the results of which are reported here, identified actors and institutions key in mediating access to benefits in this wider sense, as well as types of access and the ways that benefits moved and were shared. Attention was also paid to underlying power relations. The common themes that emerge from the comparison are tied to questions of power and asymmetries perceived by communities between their own power and that held by other actors. These themes are presented at a higher level of abstraction in this chapter in an effort to present common frames that may be more generally applicable.

Recognition

All of the common themes that arose from the comparison of the five cases of local-level discussions around benefit-sharing revolve in some way around the idea of recognition. Recognition – from both inside and outside the community – is framed as a necessary precondition for communities to secure their aims in general, including any form of benefit-sharing. In terms of power relations, recognition of the community is seen as fundamental for dialogues that might address other power asymmetries. In the Namibia case, the only one where a functioning benefit-sharing agreement is in place, this view of recognition as a precondition for progressing on other matters emerges very clearly. The Khwe community worked to produce a community protocol first and foremost because they felt that there were gaps in their official recognition as a community, and that this could potentially affect their position as recipients of benefits, or render other benefits more difficult to secure. Formal and legal recognition was also a clear element in the governance contexts in South Africa, Argentina, and Malaysia, where the community discussions described in Chapter 2 pursued different aims in contexts of relative confidence in formal recognition (flowing from, amongst other sources, national legislation on traditional knowledge and benefit-sharing in South Africa, from the Constitution in Argentina, and from an effective community association with long-standing ties with the sub-State government in Malaysia). The Greek case also confirms this view of recognition as a precondition, though this time by its absence. The traditional pastoralists felt their recognition had been lost, both informally and more formally. Recognition therefore emerges clearly from these cases as a basic requirement for any eventual benefit-sharing, or community progress more generally, to take place.

This is, of course, hardly a surprising observation – recognition is at the root of all inter-State relations, and closely linked to the principle of self-determination that lies at the core of many local and indigenous communities' struggles (e.g. Robinson, 2015). In a nutshell, if an actor is not recognized, it cannot enjoy rights. Nevertheless, the banality of this observation about the centrality of recognition is challenged by the variation in the cases. Recognition is not a uniform or straightforward theme, and manifests in different ways in the five cases, where it can be divided between a focus on actors outside the community versus those inside. Recognition from actors outside the community is a central concern in all of the case studies. Yet the focus is not only on formal (political or legal) recognition from authorities. More informal forms of recognition – of the ways of life, traditional knowledge, or cosmovisions of communities – also emerge as key. Recognition within communities is also a central theme, and often concerns the value and voice given to different groups within communities. This internal recognition of diversity proved key to avoiding perceptions of elite capture and for drafting effective community protocols (this is discussed in more depth in Chapter 5: The view from below – community protocols and international participation). The case studies provide rich examples to expand on this discussion of recognition.

Recognition from without

Beginning with recognition from external actors, a general overview of the five cases showcases the range of actors that communities, engaged in discussions around benefit-sharing, are interested in soliciting. In Greece, external actors included a range of official actors: municipal, national, and – to some extent – European Union authorities. In South Africa official actors of different kinds were of central interest – the Ministry of Health, municipal and traditional authorities, park management authorities, and private landowners. An official focus also characterizes the Namibia case, where recognition was mostly sought from national government actors. In Argentina, the focus shifts towards the private sector – though the community enjoys official recognition at the federal government level, practically useful recognition was sought at the more local level, particularly from mining companies, as well as provincial authorities. The Malaysia case also points to the private sector, as groups within the community sought more recognition from the farming company (as well as recognition from within the community itself). Nevertheless, one of the functions of the Bario food festival is arguably also to maintain the recognition of the Kelabit community amongst key sub-State government actors.

The wide range of actors communities seek recognition from is only one aspect of this question. Which actors communities 'target' when seeking recognition depends on each specific context: the lived experiences of recognition and how far these differ from formal statuses are central to understanding these choices. The Argentina case provides a useful example to explore the issue of differences between formal legal and political recognition and everyday practical implementation.[1] Argentina's Constitution recognizes its provinces as pre-existing the current federal State, and provincial authorities are thus in charge of all areas that are not explicitly assigned to the federal government – including natural resources. In 1994, an amendment to the Constitution also recognized the ethnic and cultural pre-existence of indigenous peoples and their rights of ownership over their lands,[2] as well as the right to a healthy environment. These changes fed, in different ways, into new rules about environmental impact assessments (EIAs) and the participation and consent of local communities as either local stakeholders or landowners. In addition to this, Argentina also recognizes the supremacy of international law on human rights, and has ratified the International Labour Organization's Convention 169 on free, prior, and informed consent (International Labour Organization, 1989). This complex situation regarding formal recognition (discussed further in the section on Complex governance contexts), contained in several layers of government and law, did not however translate into effective community participation in EIA processes linked to mining and exploration permits. Different local communities felt that their rights were not truly respected by mining companies, and that their permission and consent were not properly sought. The gap between recognition 'on paper' and in practice is clear in this case. It can be seen as one reason for the communities' focus on securing recognition from mining companies, discussed a great deal more than recognition from any government authority. Simply put, the mining companies were perceived as having more immediate power over local land and resources and within EIA processes than provincial or other authorities.

The point is also borne out by other cases. In the Greek case, for example, recognition was sought from a range of authorities. Essentially, the traditional pastoralists of Ikaria saw missing services as lying at the root of many of the issues under discussion. By gaining recognition from municipal authorities, services including a slaughterhouse and clarified rules for access to grazing land would perhaps follow. By gaining recognition from national authorities, land designation and ownership issues might be resolved, while recognition from the EU would allow clarity on rules applied to CAP subsidies. In South Africa, park authorities were identified, with the help of different supporting groups, as prime 'targets' for recognition by the traditional health practitioners, mainly because of their clearly defined obligations to share benefits with neighboring communities (Swemmer *et al.*, 2015). Overall, seeking recognition from formal actors outside the community emerges as central to any future scenario of benefit-sharing, and for securing community aims more generally, echoing existing literature (Wynberg and Hauck, 2014; Robinson, 2015). However, to understand which actors communities seek recognition from and why requires a closer look at specific contexts. There is no simple formula where local or national government recognition can guarantee a good setting for communities to achieve their aims, including benefit-sharing. Rather, attention should be paid to the actual lived experiences of recognition in communities beyond any formal legal or political situation.

More informal or general types of recognition, in the sense that the recognition sought was not tied to accessing a specific benefit or forum, also emerged as an important aspect of the recognition themes from the five community case studies. This can be summarized by the idea of respect for a community. The common aspects where communities sought respect were linked to their complexity and to their ways of life and cosmovisions or worldviews. Once again, the power dimension is clear: respect for a community, as a complex unit with a particular way of life and values, is a necessary departure point for dialogue between equals. This is a point discussed in the literature on deliberative democracy (Young, 2004; see also Chapter 6: Local communities, participation, and democratic environmental governance) as well as the core of good faith in scholarship on international environmental law (Morgera, 2016).

The question of the recognition of communities as complex units cuts across that concerning recognition within the community. With reference to external actors, the cases showed that this was often tied in with issues of perceived 'elite capture' – situations where external actors seek some type of consent from a single or small numbers of community members, often in exchange for payment or other goods. This phenomenon is discussed as local 'recapture' in the scholarship on CBNRM, where it is seen as one sign that power has not been effectively devolved to the local level (Murphree, 2004). Yet elite capture could also be seen as a misunderstanding of a community as a homogenous unit. This suggests a view of a community, and particularly traditional or indigenous communities, as somehow automatically sharing the same positions and interests, rather than groups that have complex internal debates and decision-making processes like any other group (this is discussed further with reference to community organization later in the chapter). That a community can comprise people with contrasting ideas and interests as well as complex decision-making processes is what emerges from the cases as an important focus for recognition and respect. Similar findings can again be found in the literature on CBNRM. Murphree (2004), for example, suggests that this kind of complex communal governance has emerged in many parts of Africa because '[g]overnments have not effectively penetrated downward into this landscape', and where they have this has not been accompanied by proper recognition, making these 'incursions [...] proscriptive, unenforceable and frequently appropriative' (209–210). Proper recognition of complex communal governance is thus inextricably linked to questions of substantive (as opposed to representative) democracy via real devolution of power in his view (ibid.).

Similar points emerge from the cases presented here. Argentina is again the most pertinent example. Questions of elite capture arose around community consent within EIA processes. In the view of the members of the Colectivo Apacheta, the simple majority procedure followed in EIA processes silenced those within the affected communities that held different opinions. The communities around Salinas Grandes were equally preoccupied with questions of consent and with avoiding elite capture, the core theme addressed in their community protocol, which explained the complex structures and timings required by their communal governance system. The question of respect and community complexity was also expressed with reference to worldviews or cosmovisions. In the Salinas Grandes communities this was made crystal clear in their community protocol.

> As we belong to Andean cultures, we seek the "Buen Vivir" before strategic development or individualist economic growth. This can only be possible if there is a harmonious relationship with the Pachamama, if participation is strengthened, if identity and cultural diversity is protected, recuperating control of our territories and lives. Our model seeks to create conditions to have a harmonious and happy life. From this logic partial visions are not possible, only integral ones. The "Pacha" is space, time, matter, spirit and also transcendental projections. Also, human life makes sense to the extent that it develops together with the others.
>
> (Comunidades Indígenas de las Salinas Grandes y Laguna de Guayatayoc de Jujuy y Salta, 2015)

As the basis for their community, this cosmovision underpins the community's complex governance models, but also underlines another aspect of complexity – the difference between the worldviews that drive external actors (characterized in the critical literature as a capitalist and extractivist conception of progress or development – see Chapter 1: Local communities and global environmental governance) and those that drive the community. The theme of respect for difference and complexity also emerges across the other cases, for example through references to *Ubuntu* as the basis for the common pooling of traditional knowledge in the Kukula Association of Traditional Health Practitioners, or in the underscoring of important traditional knowledge linked to hunting and gathering in the Khwe communities of Bwabwata National Park in Namibia, or the strong sense of duty to the community among the Kelabit in Borneo. Even in the European case, where the hegemonic worldview based in capitalism might be assumed to be the overriding norm with regards to natural resources, a similar point emerges. In Ikaria, informal recognition or respect was described as a return to the past (though this past was not tied to a specific time, appearing rather as collective memory or previously lived experience). In this past, pastoralists were respected as vital members of the local community playing a key role in society and culture, solving collective problems of water, fire control and food provision, and participating in the communal resolution of other issues, such as avoiding damage to agriculture by goats. The key point is that respect was sought in particular for a previously functional and complex system of communal governance.

Overall, informal recognition, or respect for complexity – including diverse cosmovisions or worldviews – emerges as a clear dimension of importance in the cases. It also underlines the ways in which recognition speaks to the resolution of power asymmetries, and ultimately to substantive participatory democracy – a theme returned to in more depth in the conclusions of this book.

Recognition from within

The question of the recognition of complexity and heterogeneity within communities also concerns internal interaction. In other words, recognition is a theme that also applies to relations among different sections of a single community, though in the cases here this is always in addition to questions of recognition from external actors. In Argentina, the Salinas Grandes communities intended to address how to engage more young people in their work after the community protocol had been drafted. In Ikaria, the discussion about which groups were respected players in the local community and which were not was the fundamental question at stake – some traditional pastoralists felt they were no longer included in this definition. In Namibia this theme emerged in the discussions about the mooted TEKOA school, and about how to attract local youth to learn about traditional lifestyles and knowledge. In South Africa, the possibility of formulating a code of conduct to better identify and distinguish Kukula traditional health practitioners from others was discussed in 2015. The idea was taken up more directly in later discussions about the possibility of using land recently acquired by the Association to host a training center or school for aspiring traditional practitioners.

Potential issues around questions of internal recognition were perhaps clearest in Malaysia. This case also illustrates how internal recognition may have effects on external recognition, and thus eventual benefit-sharing. In Bario, the local association, the RKS, acted as a negotiating partner to conclude the agreement for joint rice agriculture management. However, as the rice-growing project progressed a number of different actors in the community raised issues linked to farming techniques. Though not explicitly stated by community members, who expressed great trust in the RKS, the way that this group negotiated the agricultural deal does appear to play a clear role with regard to the types of issues raised, and sometimes resolved, further along in the project. The RKS negotiated a widely appreciated deal, but this did not include guidelines or rules about traditional farming practices that revealed themselves to be important for a range of reasons as the deal came to be implemented. Had there been more intra-community debate among the various different social groups within the Bario community, and particularly with farmers, these issues might have been identified at an earlier stage and built into the agricultural deal. This points to the importance of inclusive debate within communities as a criterion for effective communication with external actors. This applies even where communities already enjoy mutual respect.

This leads in turn to a final observation about internal recognition, which will be explored in relation to the common theme of organization later in this chapter. The case studies suggest that questions of internal recognition may be more explicitly or easily discussed (though not necessarily resolved) within communities of interest (or *Gesellschaft*). This is illustrated by the case of the Kukula Association in South Africa. For the Kukula, forming an Association of Traditional Health Practitioners necessarily involved explicit decisions about how to define the community, which spanned a number of ethnic and indeed linguistic groups. The Kukula Association includes practitioners that specialize in different ailments or areas of knowledge, but have decided to create an Association to pool their knowledge in an effort to both protect and perpetuate it (Cocchiaro *et al.*, 2014). Sharing their knowledge with others is not a condition of being part of the Association – rather they have pooled knowledge in order to regulate access according to common rules, including rules about training (ibid.). Thus, traditional knowledge and a shared idea of how it should be protected and passed on is the basis of membership in the Kukula Association community.

Of course, this is not a claim to any generalizable 'pattern', given the small amount of cases investigated here. Nevertheless, the idea does find some confirmation in those cases where communities are defined on an ethnic basis (or *Gemeinschaft*). For the Khwe in Namibia, the Kelabit in Malaysia, and the Atacama communities in Argentina, questions about the definition of the community were not explicitly discussed, at least when compared to the Kukula. Being a member of the community, and sharing its general cosmovision or worldview, was to some extent a given in these cases. This is not to suggest hard and fast differences – all of the cases included difficult and prolonged discussions around the identity of the community and its component parts. The observation is rather that these appear to have been raised earlier, and to some extent resolved in explicit ways, in the South Africa case of a community of interest as compared to cases involving communities of ethnicity.

Recognition plays a central role in communities' discussions of themes linked to benefit-sharing. It also forms a central node amongst the common themes and issue areas that emerged from the case comparison. In the remainder of the chapter, these other common themes – most of which have already been touched upon with reference to recognition – are explored in more depth.

Support from non-state actors as brokers

All of the cases saw some level of support provided to communities by a range of non-state (and in one case State) actors, regardless of how advanced discussions around issues of benefit-sharing were. To some extent, of course, this is due to the way that the cases were chosen. As described in Chapter 1 (Comparative local case studies), the cases were identified alongside partner NGOs. Nevertheless, a range of other non-state actors also played roles in supporting the communities in different ways. In South Africa, the Kruger to Canyons Biosphere and staff at the Rural Facility of the University of the Witwatersrand supported the Kukula Association and its members alongside Natural Justice, and funding was received from the Heinrich Böll Foundation. In Namibia, one State actor, the Ministry of Environment and Tourism, provided staff support to the Khwe residents of Bwabwata National Park, as did the IRDNC and the Legal Assistance Centre, in addition to Natural Justice and the Kyaramacan Association in which Khwe members played important roles. A nearby university also had students engaged in fieldwork alongside community members at the time of the research visit. In Argentina, the communities of Salinas Grandes were supported by locally based lawyers and FARN, and received a visit from James Anaya, the UN Special Rapporteur on the rights of indigenous people, whose subsequent report underlined their points about EIA processes. In Malaysia, the lines are rather more blurred – the role of Natural Justice had been limited to a presentation on community protocols, and the community's own RKS association secured the support of the State government and the farming company.

The range of actors providing support, their geographical proximity and presence in the local communities, and the length of the relationships, seems to have some bearing (in addition to modes of organization) on the progress made in securing communities' aims. For example, the Kukula Association, which enjoyed support from a range of locally based organizations, was making progress in developing relationships with various actors, including protected area managers. That, in turn, swelled the ranks of external support available to the group. These relationships of support often pre-dated the formation of the Association and the drafting of the community protocol, and continued as the Association redrafted its protocol and defined new goals. In Namibia, though the Khwe community protocol was awaiting its official launch, the community was already in an existing benefit-sharing agreement, itself the product of longer working relationships with various supporting actors. Ikaria, on the other hand, provides some confirmation for this interpretation due to the lack of progress made towards community goals. The pastoralists in Ikaria received support from the Documentation, Research and Action Center of Ikaria. This group is locally based but limited compared to supporting actors in the other cases in terms of the range of support it can provide. This may be due to its being a more informal group, of smaller size compared to other actors studied, and because its aims include documentation and inclusive knowledge creation in addition to drafting a community protocol from the bottom up.[3] Arguably, the lack of a functioning pastoralists' association is more pertinent here, as discussed in the following section.

Securing lasting support from a wide range of actors and the extent of community organization can reasonably be argued to be interlinked and tied in turn to the ability to gain external recognition.

Long-term support from a range of non-state actors – including actors that are local and readily accessible by communities – thus appears to boost the chances for securing the different goals communities define.[4] Most of the communities discussed had limited experience of, for example, negotiations, access to decision-makers and external procedures of decision-making, with the exception of the RKS association. It is in this view that support from non-state actors with more experience in such areas can be understood to play an important role for communities seeking recognition from external actors, even if the main purpose of this recognition is simply the right to be left alone. Support from non-state actors also raises a series of questions however. Although they cannot be answered in any definitive way on the basis of the cases presented here, they are important to raise in a more general perspective. The discussion that follows is thus hypothetical. It is expanded upon in Chapter 5 (The view from below – community protocols and international participation) in relation to community protocols, as well as in the conclusions to the book (Chapter 6: Local community participation, global civil society, and deliberative democracy).

The first question concerns the importance of long-term support from non-state actors for local communities. In this vein, the question of funding emerges as particularly pertinent. Arguments about the 'NGOization' of many non-state actors that build relationships with local communities posit that these groups are obliged to pass on factors linked to accessing funds to local communities in ways that discipline them, for example obliging measurable achievements, defined timescales, and complex accounting rules (Choudry and Kapoor, 2013). These rules are required by State or international organization funders, which channel resources through larger international NGOs (INGOs) as service providers. In turn, these groups are shaped by these rules and constrained in terms of the ways that they operate. They then pass these constraints on to local communities insofar as the projects they are able to undertake with them are shorter-term projects that allow funder requirements to be fulfilled. However, this view of non-state actor support for local communities is contradicted by the cases presented here, for the simple reason that the length of the relationship between the partner NGOs and communities was a case selection criterion. While this argument does not apply to the cases investigated here, it is nevertheless an important question to raise in considering a broader discussion of themes common to local communities.

Many of the non-state actors involved with the communities in the case studies presented here that were not part of the case selection consideration were also engaged in long-term relationships however. Of these, many are locally based actors, such as the Kruger to Canyons Biosphere Reserve, lawyers in Argentina, and the community-based NGO in Namibia. In three of the cases, the partner NGO was Natural Justice, an organization chosen precisely because it is amongst those that have reflected explicitly on elements of the 'NGOization' argument and the question of how to approach community engagement for empowerment. Natural Justice works with communities using community protocols (amongst other approaches) as a tool for legal empowerment with these points in mind (Jonas, Bavikatte and Shrumm, 2010; Bavikatte, Robinson and Oliva, 2015). The findings about the important support provided by non-state actors for the cases presented here thus apply to actors in a certain type of relationship with local communities. Nevertheless, the issue of securing long-term funding persists for NGOs that seek to work in this outlook, and funders play a key role in helping or hindering long-term engagement.

One funder of Natural Justice's work, specifically in the South Africa case presented here but also elsewhere, is the German Heinrich Böll Foundation. As a representative from this organization's Cape Town office was present at the second research visit to South Africa, a discussion about long-term funding was possible, and illustrates a different approach to project funding compared to that outlined in NGOization arguments. This foundation is bound by German financial regulations and accounting rules, but follows a different approach to project funding and measuring outcomes as a 'Green Political Organization'.[5] Since a key plank of the 'NGOization' argument revolves around the depoliticization of NGOs and international NGOs (Choudry and Kapoor, 2013), the political nature of this Foundation emerges as key to its approach, based on 3-year cycles with long-standing partners and aims or outcomes that are defined in broad terms (e.g. the provision of legal expertise in the case of the Kukula Association) which allow measurement to be more flexible. In addition, the offices of the Foundation in different parts of the world are given considerable independence. In South Africa, representatives attend events organized by the groups they fund in order to build deeper relationships and develop better understandings of the work undertaken, challenges faced, and progress made, all of which feeds into a long-term working relationship and the subsequent definitions of aims.

While this example demonstrates that alternative funding models that avoid the pitfalls of 'NGOization' are possible, funding problems were by no means absent from all of the cases presented, and questions about how to assist communities to secure funds to follow their aims did arise, in particular where supporting non state actors involved with the community did not have the freedom to allocate funds as they thought fit. Funds assigned for specific projects could not, for example, be used to cover logistical costs in the long term in many cases, and communities in Argentina, South Africa, and Namibia raised this issue in particular.

There is some potential tension in the observation about the importance of long-term relationships between non-state actors and local communities however. Usually, the longer a relationship lasts, the greater the potential risk of dependence – for example on external funding or on expertise – which goes against a goal of community empowerment. As mentioned, key NGOs involved with the local communities in the cases presented here did orient their work towards community support and empowerment, with the overall goal of assisting communities to pursue their own aims. Nevertheless, this was acknowledged to be difficult to declare 'fulfilled', particularly in light of the funding needs of local community associations. On this point, the cases suggest that community organizations that pre-date support from non-state actors may be better placed for longevity. The RKS in the Malaysia case, and the community organizations of the communities of the Salinas Grandes in the Argentina case, provide good examples. As these groups were already functioning prior to the discussions investigated here, their relationships with non-state actors were based on independent decisions to request some form of support to reach aims that had already been agreed. The outlook in cases where communities sought support to draft community protocols, but had no pre-existing organization, presented a mixed record in terms of how the community organizations subsequently formed would fare once support ended. Again, the case studies only suggest points for further exploration, but the question of longevity is generally highlighted as a crucial component for collective action groups to achieve their stated aims (e.g. Bosi, Giugni and Uba, 2016).

The focus on long-term relationships that emerges from the case comparison also points towards another aspect of the dependence risk: long-term relationships might also see communities define their aims to coincide with those favored by a non-state actor. This is certainly not the intention of the non-state actors involved in the cases described here. Nevertheless, whether intended or not, this remains a possible outcome that could be argued from the 'NGOization' point of view to distort the aims and wishes of any given community. In that argument, distortions can occur through project-based funding, and in particular the definition of which projects should be funded, which acts as a disciplining force on non-state actors and in turn the communities and organizations they work with (Bayrak, Tu and Burgers, 2013). The possibility is worth raising in a discussion of the tension between community independence and long-term relationships with non-state actors, and points once more to the advantages of independence linked to pre-existing organizations in the cases explored here. A further tension with earlier findings is suggested here however. It was suggested that communities of interest, which tend to be newer organizations, were better suited to address issues of internal recognition compared to long-standing traditional organizations. The discussion of long-term relationships with supporting non-state actors and the risks of dependence and distortion partly contradict this view, to the extent that the case studies suggest newer organizations to be in more need of external support.

A clear tension is suggested by the case studies between old organizations and new, the need for community independence and support from non-state actors, and the twin risks of dependence and discipline on the one hand, and the reproduction of existing power asymmetries on the other. This is, of course, a polarized portrait, and the cases explored show that in reality these balances are struck in a variety of ways. By highlighting the tensions at play, however, we gain a better idea of the complexity of the decisions and strategic choices communities and non-state actors alike are faced with, and the importance of how non-state actors provide support is emphasized.

Community organization

Another area where delicate balances need to be struck emerges in the forms and models of community organization. One clear common point that arises from a comparison of the cases is that the more hierarchically and formally organized the community, the more progress made towards the community's stated aims. There appear to be clear knock-on effects between formal organization and external recognition, which allows access to different forums of decision-making, as well as procedural guarantees, including access to information. These knock-on effects can ultimately feed into communities fulfilling their aims. The South Africa case is a good illustration to this end. The Kukula Association was created following a classical hierarchical model, with a CEO and board. Once formed, the Association was the vehicle for drafting the Kukula community protocol, but also for increasing contacts and relationships with a range of external actors. In Malaysia, the Kelabit community of Bario also drew on a strong and formally organized association. At the time of writing, these two cases showed the most progress towards the aims the community had set themselves. The traditional pastoralists in Ikaria once again form the opposite image – there is no functioning formal association, and this case is (again, at the time of writing) the least advanced in terms of the community's goals.

Thus, the cases suggest that formal organization is useful for gaining ground in terms of the goals a community sets itself. This is unsurprising at least as far as benefit-sharing is concerned, since by definition this involves dialogue geared to securing some good from external actors. In this view, formal organization can be argued to make it easier for external actors to recognize and engage with communities. Most of the external actors that communities sought to address regarding issues linked to benefit-sharing in the cases are themselves organized in formal and hierarchical ways, and similar forms of community organization make them more familiar and thus visible and comprehensible. In South Africa, the CEO or management committee of the Kukula Association is a clear point of contact. In Greece, the lack of an association of pastoralists meant that at moments of importance to the community, such as the discussion of the land use designation map at a municipal council meeting, there was no recognizable representative able to participate on behalf of the pastoralists.

However, there is a potential mismatch between formal and hierarchical organization and different local communities that is suggested in the other cases. This concerns whether or not this type of formal and hierarchical organizational model fits with the ways that a community conceives of the world – its cosmovision or worldview – and how this translates into ways of organizing. Cosmovisions or worldviews that differ from a more western outlook (see the discussion in Chapter 1: Local communities and global environmental governance) have clear effects on organization in two of the cases in particular. In Argentina, the Salinas Grandes communities state their case for consultation procedures in line with *Buen Vivir*, and the ways in which they organize and take decisions also follow this outlook. In Namibia, the Khwe have traditionally organized in non-hierarchical and consensus-oriented ways. The election of a committee for the purposes of drafting a first version of the community protocol was not a mismatch with this tradition, since the draft was then shared with, amended, and agreed with community members in a participatory process. Rather, a potential mismatch with a formal hierarchical organizational model emerged in the discussion about the appointment of a Khwe traditional leader. One of the various reasons for which no leader had been recognized was simply that no such role had previously existed in the Khwe community, making the nomination of a leader complex. Other work focusing on the Khwe has also underlined a mismatch between hierarchical organization and traditional Khwe approaches. Specifically, the case of the benefit-sharing agreement on Hoodia studied by Vermeylen (2007) discusses problems experienced with this agreement as rooted in such a mismatch.

This point about potential mismatches between cosmologies or worldviews and forms of organization links to the observations discussed earlier in this chapter with reference to external recognition of different ways of life. Where this is expressed through different organizational forms, the capacity of external actors to recognize and accept these unfamiliar modes of organization is key for the recognition and respect of a community. This is discussed further in Chapter 5 (The view from below – community protocols and international participation) with regard to community protocols. These documents, as is clear in the Argentina case, can be valuable tools for translating community cosmovisions or worldviews for those unfamiliar with them. Yet this is arguably an unusual scenario, and the reality for many local communities is different in that they are subject to various pressures to organize in formal ways. This pressure can come from non-state actors, through the kinds of mechanisms described in NGOization arguments outlined in the previous section, from State actors, as seen in the example of the Khwe traditional leader in Namibia, or indeed through law. In this view, the question of how best to organize is another area where communities need to strike a balance between remaining close to cosmovisions and worldviews, and the models best suited to securing their aims, at least where these two factors are not aligned.

The South Africa case suggests that alignment between an organizational form suited to reaching goals with external actors and a community's cosmovision or worldview may be more likely to exist precisely where a community is new, and based on interest. The Kukula organized in a formal and hierarchical way. At the same time, because of the newness of the organization and the languages and ethnicities it spanned, the group was obliged to discuss its self-definition and decision-making mechanisms explicitly. Thus, for the Kukula, there was alignment between organizational forms dictated to some extent by the needs and requirements of external actors (including the law) and the need to ensure internal recognition. This observation is however supported by a single case, and there is thus little reason to assume it is more generally applicable. The community of interest in Ikaria, a community that has existed for a much longer time as an identifiable group, did not provide any opportunity for discussing such questions of self-recognition. This suggests that time, like the basis of a community's definition as rooted in interest or ethnicity, is just as important for an organization to be both effective and inclusive.

Other explanations thus need to be considered. Another possible reason for alignment between these two needs, this time suggested by more than one of the cases, is the purpose for which a community organizes. Is an organization founded for, or otherwise geared specifically to, the purpose of discussing issues around benefit-sharing, or is it a long-standing, perhaps traditional community governance institution that is simply attempting to deal with issues around benefit-sharing alongside other business? This unites the South Africa case of the Kukula Association, where a new organization was formed, with the Argentina case, where the communities of Salinas Grandes drew on their traditional organizations rooted in *Buen Vivir* but reconstituted them with the specific task to draft a community protocol that aimed at translating the communities' cosmovision or worldview for external actors to understand. In both of these cases, some progress towards community goals was seen. The Malyasia case bolsters this view by providing a contrasting picture. The RKS, which negotiated the agricultural deal, is a long-standing organization created to represent the Kelabit people in an array of areas, and to act as a 'bridge' between the community and the sub-State government. Negotiating issues around benefit-sharing was not an original goal of this association, nor was representation of the community in any negotiations or other decision-making forums, and this can reasonably be pointed to as a factor in the issues that arose in this case around farming techniques.

How communities organize emerges as crucial in discussions around benefit-sharing. The cases suggest several ideas that point to another potential source of tension between community values and aims. First, the cases suggest that formal and hierarchical organization is useful for securing aims linked to benefit-sharing, since power over these often belongs to external actors that find it easier to recognize and engage with communities organized in these ways. Second, how communities organize traditionally in line with their cosmovisions or worldviews may not always be in line with such formal, hierarchical models. Here, the cases provide some suggestion that alignment may be more easily achieved where new community organizations are formed specifically for addressing questions linked to benefit-sharing, or where existing organizations are geared towards communicating or translating community cosmovisions or worldviews to external actors. In cases illustrating both of these scenarios external actors made moves to recognize communities.

Complex governance contexts

While the ways in which communities organize and support they may require from non-state actors are themes more closely linked to what goes on within communities themselves, the themes of governance contexts and procedural guarantees speak to the formal contexts that communities exist in, and how these might be navigated. That local communities exist within complex legal and political contexts is a rather obvious feature to take note of in comparing the case studies. However, complex contexts shape the challenges and opportunities faced by local communities in a number of ways. In the discussion of recognition, governance contexts were identified as a key explanation for the actors singled out as important sources for external recognition. In addition, governance contexts shape the strategies and aims of communities more generally. It could again be argued that this observation is a consequence of the research design – as the research presented here departs from a constructivist standpoint, understanding the world as built in social interactions, this conclusion is already assumed (see Chapter 1: Studying benefit-sharing from the bottom up). The contexts in which discussions of issues linked to benefit-sharing are framed are automatically seen as a crucial factor in explaining the case studies. The aim of this section is not therefore to make the point that context counts, or even that contexts are complex – this is generally true and accepted in the literature. Rather, the aim is to look more closely at how these contexts shape local community discussions and actions in different ways, and to identify some features of governance contexts that help communities achieve their aims related to benefit-sharing. Clear national legal frameworks within more centralized State structures emerge as useful in this vein.

Each of the case overviews in Chapter 2 begins with a sketch of the complex governance contexts that communities exist within, and argues that these contexts are key to understanding them. The discussion here will focus on selected cases, demonstrating how each community's aims and strategies can be understood as partly shaped by governance contexts. This moves the discussion beyond the observations about recognition from external actors presented at the beginning of this chapter, though the ways that contexts shape communities are still closely connected to questions of external recognition.

Argentina's governance context, including implementation gaps and inconsistency, was outlined at the beginning of this chapter. To recap, Argentina's Constitution gives supremacy to international human rights law within its domestic legal framework. Moreover, of particular significance for the communities around the Salinas Grandes and Olaroz salt flats, in 2000 Argentina ratified ILO Convention 169, which lays down rules on consultation and consent for indigenous communities (see International Labour Organization, 1989; Marchegiani, Morgera and Parks, 2019). Amendments were made to the country's Constitution in 1994, adding the right to a healthy environment and recognition of the fact that indigenous peoples and communities existed before the foundation of the State. This led to the adoption of a new General Environmental Law in 2002 that included rules on consultation and consent, and to the transfer of land titles to many local communities now recognized as original landowners. As Argentina is a federal state, its provinces have jurisdiction over natural resources, and must apply general framework laws at the local level. In Jujuy, the province where most of the communities discussed in the Argentina case are located, the Environmental Management Unit is responsible for consent and consultations within EIA processes. In 2011, the Jujuy province declared lithium a strategic natural resource in line with federal government policies (Marchegiani, Hoglund Hellgren and Gomez, 2018), which implied an additional review of proposed projects, though the committee is arguably more likely to be in favor of permitting mining, given its strategic status (ibid.). Returning to the federal level of government, legislation on mining in Argentina includes rules about annual fees to be paid to relevant authorities, as well as 'the permission of the land "owners" to explore for minerals' (Anaya, 2012: 11). There are no rules, however, about community consultation, despite the requirements outlined in the General Environmental Law.

The complexity of this legal backdrop, including inconsistencies between different laws, is further complicated by an implementation gap at the local level (this is also discussed in the following section on procedural guarantees). In addition, the political context is one of clear support for lithium mining as a strategic resource for the country. The strategies followed by the communities located around the Salinas Grandes and Olaroz salt flats reflect this complex backdrop. In both areas communities first followed legal strategies in an attempt to ensure EIA processes were implemented correctly. Second, international legal and political strategies were also followed more explicitly than in other cases, as the communities outlined their grievances to the then UN Special Rapporteur on indigenous rights, James Anaya, in 2011, as well as to the UN's Economic Social and Cultural Rights Committee. In June 2013 they brought a complaint before the Inter-American Commission on Human Rights.

The choice to follow legal strategies in the first instance can be seen as a response to the implementation gap in national law, while the choice to appeal to the international level reflects Argentina's recognition of the supremacy of international human rights law. International strategies followed with the intention of achieving local-level effects are termed 'boomerang' strategies in the literature on activism, which depicts them as common choices where national legal frameworks afford less space for effective change compared to international regimes (Keck and Sikkink, 1998). This is arguably the case in Argentina, since more detailed requirements for community consultation and consent are in place in international sources (Marchegiani, Morgera and Parks, 2019). The Salinas Grandes community protocol can also be read as a response to the complex governance backdrop in Argentina. The protocol provides detail about how community consultation and consent procedures should look, outlining a procedure that respects the communities' worldviews and decision-making processes.

The interlinkages between international and domestic levels of governance also play a clear role in shaping community actions in Ikaria. At the international level, the importance of EU legislation for the everyday livelihoods of pastoralists derives from the CAP and food safety and animal welfare legislation. Initial increases in the numbers of goats in Ikaria were partly attributed to the CAP, while difficulties in managing the rising animal numbers were attributed to challenges in complying with food safety and animal transport legislation in the context of a small island with no approved slaughterhouse. The CAP was also identified as a source of confusion among pastoralists, with complex rules on subsidies that changed over time with periodic reforms of the policy. In addition, the EU's role in Greece's financial crisis was named as one explanation for the perception of the central government as 'grabbing' land by designating industrial zones on a land use map. In this interpretation, requirements attached to financial aid provided to Greece by the EU (and other international actors) were viewed as obliging the central government to raise revenues from all possible quarters, including by liberalizing various service sectors. The land designated for industrial use on the map was linked by some community members to preliminary approval for the construction of a wind turbine park granted by the Greek Renewable Energy Authority, thus linking the map to the generation of rents for the State. Complexities in national law also contributed to the confusion that surrounded the land use map. Various community actors described their confusion over Greek laws that they felt ultimately contributed to the use of the map to allow the central government to claim land. These laws included those on land inheritance, rules about how to prove land ownership, rules for land designated as forest or for reforestation, and accompanying rules prohibiting animal grazing (see Tsioumani, 2020, for further details).

The complexity and uncertainty that characterized pastoralists' and other community members' views of the governance backdrop in Ikaria is one source for explaining the comparatively low level of action in this case. Pastoralists were faced with a complex situation in which to find an effective course of action that would allow them to return to their status as a respected group in local society (which they felt they held in the past but had since lost). In addition, the daily challenges experienced in this governance context left little time for pastoralists to organize. Working to understand and comply with EU rules in order to secure vital subsidies, in addition to normal workloads, made collective action particularly challenging.

Similar arguments can be made about the other case studies, though the Argentina and Greece cases are perhaps the most emblematic. In Namibia, concerns about the status of the Khwe communities resident in Bwabwata National Park underpinned the decision to draft a community protocol, despite their clear and official role in aspects of park management via the Kyaramacan Association. In South Africa, the Kukula Association's main aims revolved around recognition from external actors, defined by the institutional set-up in the country with its complex and overlapping attribution of competences to traditional, municipal, provincial, and national authorities as well as protected area managers. In Malaysia, the decision to request subsidies from the sub-State government followed other attempts to boost rice agriculture through different international and national schemes, and depended on the particular form of community NGO in the country.

Overall, though an apparently banal observation, pointing out that community actions are shaped by increasingly complex governance contexts underlines the importance of considering the unique circumstances each community exists within. Attention to local specificities is necessary to paint a complete picture for each case. This general finding is discussed at length in the wider literature on activism and social movements as political process or political opportunity theory, which argues that different types of activism should be read as deliberate and considered reactions to complex governance (Meyer, 2004). Because complex governance contexts are common to all communities, attention to unique contexts suggests that 'one-size-fits-all' policies for local-level benefit-sharing, or indeed other subjects, are unlikely to lead to optimum results. This chimes with the findings of a great deal of literature, including that on polycentric governance for common pool resources (see e.g. Ostrom, 2010) and CBNRM (Murphree, 2004; Nelson, 2010).

More specific ideas about what type of governance context might be conducive to eventual benefit-sharing are also suggested by the case studies. In particular, clear national legal frameworks for benefit-sharing appear to play a useful role, even where there is uncertainty in other areas of governance. South Africa provides the example here, with its national legal framework that supports benefit-sharing. This is made up of, for example, clear mandates for benefit-sharing from protected areas, a legal process for land claims leading to other forms of benefit-sharing such as co-management, and national legislation on benefit-sharing from bioprospecting involving indigenous biological resources (Biodiversity Act No 10 of 2004), which implements the Nagoya Protocol to the CBD (Convention on Biological Diversity, 2010). The existence of these provisions appeared to have clear effects for the Kukula Association, since decision-makers and managers had clear notions of what was required in their relations with communities, and resources were also available in some cases to fulfill their mandate in that regard. Protected area managers in particular were fully aware of their duties to engage and benefit the communities surrounding protected areas under South African law, and were thus prepared to engage in dialogues to find solutions to the issues raised by the Kukula Association.

Although South Africa is the only case where the legal framework around benefit-sharing was developed to this extent (at least during the research period; Namibia has since adopted legislation implementing the Nagoya Protocol to the CBD (ibid.)), other cases provide some confirmation of this idea about the helpfulness of clear legal frameworks by virtue of their absence. The only partial exception is Malaysia, where a legal framework is in place, but only for benefit-sharing from genetic resources. This suggests that benefit-sharing needs to be considered across policy and issue areas to better aid local communities in complex contexts. The lack of a legal framework on benefit-sharing can be understood to constitute a barrier to communities in a comparative perspective, since external actors in other cases simply did not have the knowledge of benefit-sharing noted in the South African context.

The five case studies suggest that complex governance backdrops add to the challenges faced by local communities, albeit in ways that can only be understood by paying attention to specific contexts and the perceptions of governance held amongst community members. One more specific observation, based in the main on the South Africa case, is that clear legal frameworks on benefit-sharing can play a role in reducing confusion in complex contexts where they assign unambiguous roles to external actors that communities must interact with in issues linked to benefit-sharing. Such a framework was only in place in South Africa. It is thus worth looking at other features in the governance contexts of local communities that may help navigate issues linked to benefit-sharing. Some possibilities related to procedural guarantees are discussed in the next section.

Procedural guarantees (information and voice)

Procedural guarantees refer to what communities expected various actors to provide in terms of the basic services and spaces necessary to allow them to make informed decisions in complex contexts. They complement the findings on complex governance backdrops by outlining further factors (in addition to clear and wide-ranging legal frameworks on benefit-sharing) that can help communities to navigate complex governance contexts. The label 'procedural guarantees' is an abstraction used to describe a range of case-specific experiences, which point to these helpful factors. These experiences can in turn be grouped according to: 1) access to information that can be understood by community members (i.e. is accessible and comprehensible) and is perceived as neutral and reliable, and; 2) clear and accessible forums for expressing opinions and concerns around benefit-sharing issues, both between communities and various external actors, and for intra-community discussion. Recognition is, as discussed earlier, a clear prerequisite for the proper functioning of these procedural guarantees – without respect and recognition, communities' voices may be expressed in different forums and information supplied, but little may be done to act upon them. Nevertheless, fulfilling these procedural guarantees was seen as necessary for any future benefit-sharing scenario in the cases presented here.

The Argentina case is the most informative on the issue of access to neutral, reliable information in an appropriate format. Both of the groups discussed (the communities around the salt flats of Salinas Grandes and Olaroz) raised concerns about information. Despite clear obligations flowing from international, national, and provincial law, the communities felt that information was either missing, provided reluctantly or in an untimely manner (leaving too little time for proper consideration), provided in formats ill-suited to most community members due to highly technical language, or biased. The community members of the Colectivo Apacheta raised some of these points. In line with the EIA processes outlined in national and provincial legislation, in Olaroz companies contacted local authorities, which take the basic form of a local executive composed of an elected president and committee acting in accordance with decisions made by a local assembly where all local residents vote – given the small sizes of the communities this is possible (Marchegiani, Morgera and Parks, 2019).

However, the community members opposed to mining underlined that information about these projects was not provided early enough for them to read, understand, and consider before community assemblies were held, and that the presence of company representatives who could answer questions about the projects was not always communicated in advance. They also noted that the information provided was in the form of long technical reports in hard copy, held by elected representatives for them to consult. Company representatives also used technical language in their presentations and discussions at local meetings. Community members were, in addition, concerned that the information provided was from the mining companies rather than independent sources. This observation also ties in with the gap between implementation capacities and legislation noted earlier with reference to recognition from external actors: in this case the provincial government did not have the resources necessary to commission scientific studies. Though even if it had, suspicions were also raised with reference to this body, which some saw as prioritizing mining over local community rights and interests given that lithium had been declared a strategic resource.

The community members that raised these points in the Olaroz area were worried above all else about the possible contamination of closed water basins, the only source of fresh water for the communities, and their livestock, in this very arid part of the planet.[6] Though they acknowledged that other community members felt differently and welcomed the mining projects due to the investments and jobs they would bring to the area, these community members remained adamant that the information requirements of the EIA process had not been fulfilled adequately. The communities in the Salinas Grandes area shared these concerns. Here, however, no mining was underway and the decision was taken to draft a community protocol that would clarify the process of community consultation for external actors, with particular attention to questions of information. Indeed, the protocol underlines the need for independent information in accessible formats (with decisions about formats left to the community), and for meetings to be organized well in advance, in local and accessible places, and at appropriate times. It also calls for experts to be available to communities to aid them in interpreting technical information and for good-faith dialogue with external actors more broadly (Comunidades Indígenas de las Salinas Grandes y Laguna de Guayatayoc de Jujuy y Salta, 2015).

The case of lithium mining in Argentina thus revolves around wider questions of information and consultation. Similar issues also emerge from other cases. Information, or rather the lack of clear information, was a key issue in the disputed land use map in Ikaria. A similar lack of clear information about the local community's status was highlighted in the case of Khwe residents of Bwabwata National Park. In Malaysia, a lack of information about the farming techniques to be employed in the rice cultivation project was identified as a possible reason for issues that emerged in its implementation. In contrast, clear information about the duty of protected area managers in facilitating benefit-sharing with neighboring communities was central to the progress made by the Kukula Association in South Africa in terms of their relationships with protected area managers. The theme of information and its importance also features prominently in international guidance about local community consent and benefit-sharing related procedures, including, for example, recent guidelines from the CBD (Convention on Biological Diversity, 2016), which were drafted with input from indigenous peoples' representatives (Parks and Morgera, 2015; Morgera, 2017).

The second key element to emerge from the case studies under the rubric of procedural guarantees was the need for identifiable and accessible forums in which communities could express their opinions and concerns. Calls for these spaces were often linked to communication with external actors, though similar needs for intra-community dialogue also emerged in some cases. Identifying forums for dialogue, for example, forms a key part of the Salinas Grandes community protocol (Comunidades Indígenas de las Salinas Grandes y Laguna de Guayatayoc de Jujuy y Salta, 2015). In other cases, spaces for dialogue with external actors were not clearly defined, but their necessity was discussed in more or less explicit ways. In South Africa, for example, the Kukula Association made progress in meeting with external actors at periodic Association events and on other occasions, but these were sporadic. The view that a more regular meeting forum was necessary arguably underpins comments concerning the lack of feedback following interactions with external actors mentioned by some members. In Greece, the view of the past as a sort of ideal model for the solution of issues in the present also includes considerations about spaces for dialogue, though this time intra-community in scope. In particular, pastoralists and other community members described the communal resolution of a range of local issues through village meetings. These meetings were used to resolve problems and decide on divisions of labor, with everyone accepting the solutions agreed. A clear example was the system of communally maintained gates and walls designed to keep goats off agricultural land.

While other cases underline this general point, the Malaysia case points specifically to the need for clearly identified spaces for dialogue. Issues raised by community members about farming techniques in the agricultural project underway in Bario were raised in a range of different arenas: via block leaders with local managers employed by the farming company, directly with local managers, with farming company managers based in the city of Kuching, and via the RKS community association. In this case, the number of spaces available for raising issues contributed to uncertainty – both about which issues to raise in which forum, and about the extent to which different community actors were aware of and responding to the issues raised. The existence of too many spaces for dialogue thus also appears as problematic, though less serious than the absence of such spaces.

What emerges from these examples drawn from the cases is the idea that clearly identifiable and accessible spaces for dialogue are an important part of the equation when local communities discuss issues linked to benefit-sharing. These spaces are the more concrete accompaniment to recognition from without and within the community – they are the spaces where dialogue allows recognition to take place. Combined with access to neutral and timely information, these procedural guarantees suggest practical elements to allow for recognition to be constructed amongst communities and a variety of external actors via dialogue and shared knowledge.

Non-monetary benefits

The remaining common themes to emerge from a comparison of the five cases presented in Chapter 2 concern the ways that communities frame benefit-sharing. The first common frame, though it is expressed in different ways and with reference to very different contexts and concerns, is that non-monetary benefits are consistently viewed as of equal and sometimes higher importance than monetary benefits (though these were by no means seen as unimportant). This is explicitly linked to the idea of recognition through the view that communities should have a voice in defining benefits (and thus ideas of what is 'good' or what 'development' or 'progress' is). This point is also discussed in the following section on the second common frame of benefit-sharing as a process.

In Malaysia, the Bario community's agricultural deal provides a first example of the ways in which non-monetary benefits are stressed. The agricultural deal for boosting Bario rice cultivation revolves around monetary benefits in the form of State subsidies. Yet the central questions raised about the implementation of the project stressed non-monetary aspects about the most appropriate farming methods to cultivate Bario rice in a sustainable way rather than monetary questions around profit-sharing. The framing of benefit among those farmers that raised these issues concerned the long-term future of the cultivation of Bario rice for the community. The benefits they called for relate to the resumption of more traditional farming techniques long used in the area, and the conservation of the quality of Bario rice. The central benefit was thus rice itself, as well as its importance and significance for the Kelabit way of life. Few of the concerns about cultivation were linked by local community members to aspects of monetary benefit such as yield or profit margins. Rather, they underlined respect for traditional knowledge and the unique qualities of Bario rice. This relative lack of attention to monetary benefit-sharing could be argued to be logical since the agricultural deal had already secured State subsidies used to fund various infrastructure improvements for local farming as well as the services of the farming company for agricultural services. Yet monetary questions were not altogether missing from discussions in Bario. The emphasis on non-monetary benefits may also be linked to Kelabit culture, though a real understanding of this aspect is beyond the scope of this work. It is also likely linked to the high levels of trust in the RKS, which is entirely made up of Kelabit community members who feel a strong duty to their community, wherever they may reside.

Namibia is another case where benefit-sharing that is mainly monetary or economic in nature is already in place, via the Khwe community's involvement in the Kyaramacan Association. The Khwe framing of benefit, at least in the context of their community protocol, nevertheless emphasizes the importance of the possibility to protect and transmit their culture and traditional knowledge. This cultural reproduction is the central non-monetary benefit that various other aims revolve around – for example legal representation and certainty, access to different areas of the park, and the foundation of a school. The Khwe community felt that cultural reproduction would allow them to feel more secure in their position as residents of the park – to feel recognized, in the language of this analysis – and able to work on reviving and guaranteeing the future of their community. Again, existing monetary benefit-sharing could be argued to motivate this focus on non-monetary benefits in the community protocol. Yet once again monetary benefits were not entirely missing from the debate, indicating that the emphasis on non-monetary, culturally rooted benefits was a matter of choice. The reasoning behind the inclusion of more monetary aspects of benefit in the community protocol came from the Ministry of Environment and Tourism. Advice from this quarter to the Khwe community was to link the protocol to clear international sources (such as Article 8(j) of the CBD and the Nagoya Protocol to the same), but also to phrase aims in ways that would tie in with national development goals, for example by referring to ecosystems services.[7]

An emphasis on non-monetary benefits, as well as the link between non-monetary benefits and the recognition of community worldviews or cosmovisions, emerges more strongly in the other cases, where no monetary benefit-sharing is in place. In Argentina, the discussions among the communities of the Salinas Grandes and Olaroz salt flats ultimately revolved around the meaning of 'development' – seen as a clash between definitions rooted in economic growth and *Buen Vivir*. The Salinas Grandes community protocol makes this point clearly. It argues that the large-scale extraction of lithium from the salt flats, as well as other extractive activities in the area, 'value technology and despise labour power', and explains that their opposition is not to mining per se: 'We do not resist the use of technological tools, provided that they support the extraction of salt in a rational way.' What is 'rational', in this view, is defined not in economic terms, but with reference to respect for the earth and for the broader community – for salt as a 'living being' (Comunidades Indígenas de las Salinas Grandes y Laguna de Guayatayoc de Jujuy y Salta, 2015). The centrality of non-monetary benefit, defined here as respect for traditional balances between communities comprising people and other living beings including the earth, is explicitly linked to a worldview in this case, and lies in clear contrast with the monetary benefit-sharing focus in national legislation. The country's Mining Code, for example, contains rules on benefit-sharing only in terms of the payment of annual fees.

In South Africa, the theme of cultural reproduction – that is conserving and passing on traditional knowledge and skills – emerged through discussions about training, recalling the experiences of the Khwe in Namibia. The practical aims of the Kukula Association, which were at the time of the research geared towards obtaining official recognition and thereby access to the medicinal plants necessary to their work, are ultimately concerned with this non-monetary benefit. As already noted, the first version of the Kukula community protocol, as well as the formation of the Association itself, were intended to create a common pool of traditional knowledge in order to regulate access and protect this resource for future generations. In Greece, the aims of traditional pastoralists were linked to a return to a time when communal decision-making took place with fewer intrusions from external governance structures. Overall, these cases reinforce the point about attention to non-monetary benefits, mostly framed in terms of protecting resources for future generations. This protection requires 'benefits' that are defined in different ways – respect for natural resources even when extracting them, access to plants including through the protection and rejuvenation of lands outside protected areas, or returning to local forms of decision-making. But all underline the common theme and a clear link to local cultures and values.

Some degree of contrast emerges between the few national instruments hinted at in the case studies and this emphasis on non-monetary benefit-sharing among local communities. The benefit-sharing agreement in Namibia is not only monetary, but benefits are economic rather than the cultural benefits stressed by communities. The agricultural deal in Malaysia also focuses on profit- and product-sharing with local community members, who in turn were more concerned about farming practices. Discussing benefit-sharing in existing international instruments, Morgera (2016) distinguishes between inter- (among States) and intra-State (within State) benefit-sharing. The case studies presented here discuss issues related to the latter. In Morgera's view, such intra-State benefit-sharing must comprise the co-development of definitions of benefit, as well as benefits themselves. It is this component that the case studies suggest remains underdeveloped, most tellingly through the emergence of recognition as a key central theme.

Intra-State benefit-sharing and the emphasis on monetary views have also been discussed at the international level. James Anaya, the former UN Special Rapporteur on the rights of indigenous peoples, argues that concepts of benefit-sharing should move 'beyond restrictive approaches based solely on financial payments which, depending on the specific circumstances, may not be adequate for the communities receiving them' (Anaya, 2010: para. 80). Scholars also argue that monetary benefits can cause negative outcomes where elite capture takes place (Nelson, 2010; Gilbert and Doyle, 2011), and link this to a lack of true engagement and understanding of the communities concerned (Murphree, 2004; Vermeylen, 2017). There have been moves at the international level within the CBD to address this – the Akwé: Kon Guidelines for Cultural, Environmental and Social Impact Assessments, for example, note that benefit-sharing should extend beyond monetary payments to include access to markets and the diversification of economic opportunities for local businesses (Convention on Biological Diversity, 2004: para. 46). This still retains a wider economic emphasis however.

There is some evidence here that points to more monetary, or at least economic, views of 'benefit' prevailing in existing instruments. Non-monetary definitions that emphasize cultural factors instead emerge forcefully from the case comparison. This points to the idea that measures to develop the quality of dialogue – such as those discussed in the section on procedural guarantees – are key to bridging this gap. The local communities studied also suggest this: the final common theme is precisely that benefit-sharing should be seen as a process, rather than as an event.

Benefit-sharing as a process

The final common theme to emerge from the case studies is that communities do not frame benefit-sharing as an outcome, event, or contract, but as a continuing process of mutual understanding and dialogue, where definitions of benefit necessarily change over time as community needs evolve. All of the case studies point to this framing of benefit-sharing as a long-term, partnership-building process, rather than an event or a one-off fix. This view points to a conception of fair and equitable benefit-sharing, which must be based on dialogue and processes, geared towards mutual understanding. These dialogues refer both to actors within communities, who must build shared expectations, community aims and definitions of benefit, and to actors external to communities, who should engage with these community-based aims and definitions. Understanding benefit-sharing as an ongoing process also underlines that communities are complex and changing units. Their needs and their definitions of benefit will necessarily evolve over time. A view of benefit-sharing as an ongoing and open-ended process is better suited to accommodating this.

The case studies suggest this framing in a number of ways. The first element that points to benefit-sharing as a process is time. All of the cases demonstrate that time is a crucial factor in community discussions related to benefit-sharing. Most of the communities studied did not discuss benefit-sharing directly, concentrating instead on what could be argued to be preconditions. Although the cases were selected with the progress in discussions linked to benefit-sharing in mind, timeframes for developing shared understandings around benefit-sharing issues proved lengthy regardless. Whether discussions were at earlier stages, as in Greece, or had progressed as far as a revision of an existing community protocol, as in South Africa, the time required for benefit-sharing dialogues both within communities and with actors outside the community emerged as key.

Time in turn underlines the idea of continuous negotiations of shared understandings linked to benefit-sharing within communities. The Salinas Grandes community protocol provides a good example in this regard. Although the Salinas Grandes protocol refers mostly to consultation processes and information requirements, and the communities were unwilling to discuss any form of possible benefit-sharing since this risked implying eventual consent to mining or other projects, the community protocol provides a view of how continuous community dialogue and fair and equitable benefit-sharing are linked. The protocol focuses on what consultation and consent meant to the communities, and compares these with the natural lifecycles of the salt flats. This paints a picture of consultation as a process, but also as an ongoing, circular process of dialogue that takes account of change and involves efforts to understand how change is unfolding. A similar point is also reflected in other case studies. In Greece, for example, the need for dialogue between sections of the community to better understand the factors contributing to local problems was a strong theme. In South Africa, community discussions regarded the community protocol as a living document to be updated periodically to reflect changing needs and aims.

Regarding ongoing processes of dialogue with external actors, the case studies also show that time is a major factor when beginning and pursuing dialogues with external actors. Here, the cases underline that ongoing dialogue over time is required to develop the trust and understanding required for meaningful dialogue. The Kukula Association in South Africa is a good example in this vein. The Kukula began their external dialogues with an application for access to the Mariepskop Forest Reserve, supported by various non-state actors. A range of different possible solutions emerged from this process as staff from the Forest Reserve came to understand the community's needs, and as the Kukula members came to understand the constraints and issues linked to plant conservation and care faced in turn by the staff of the Forest Reserve. At the outset, the Kukula members sought access to the reserve to harvest plants in sustainable ways, which proved challenging, particularly in terms of logistical organization. A different solution thus emerged, where community members could contribute to the Forest Reserve's mission of improving biodiversity outside the protected area by growing and caring for plants in their local area. This in turn led the Association to begin dialogues with traditional and municipal authorities to access land. Once again, the other cases point to similar ideas, whether dialogues were an aim for the future (traditional pastoralists in Ikaria), well underway (the Khwe in Namibia), or ongoing (the Kelabit in Malaysia).

All the cases point to dialogue as a basis for building trust between communities and other actors. It is in dialogue that the transparency that flows from the types of procedural guarantees discussed earlier ties in with this overall framing of benefit-sharing as a process. For trust to be built, and recognition to be secured and meaningful, communities must have guaranteed channels for information and spaces to express their voices. These are key elements to building benefit-sharing processes that may end in trust (or at least good faith, to use legal terminology). This point is underlined by the case of the agricultural deal in Malaysia. Here, the community's trust in its RKS association led to an effective deal to boost agriculture in the area. However, the procedural guarantees for information and voice were arguably not guaranteed despite widespread trust – there was as a result some confusion about where issues should be raised, given the existence of a number of different forums for voice, while information about the agricultural deal did not include details on farming techniques that were later raised by some community members.

The framing of benefit-sharing as a process is practical in the sense that it informs us about the concrete aspects needed. That achieving eventual benefit-sharing will take time, trust, and procedural safeguards to allow actors with potentially diverging interests and outlooks to come to meaningful understandings and agreements about whether and how natural resources in local communities should be used is hardly surprising. Indeed, existing literature on benefit-sharing agreements in local communities paints a similar picture (Robinson, 2015). The clear link between preconditions for processes of benefit-sharing, such as rules for consultations and consent, is not reproduced in the case studies. Both aspects are considered, but not in holistic ways – rules about consultation and consent see these as processes, while benefit-sharing is conceived of as a contract. The same image appears at the international level. The Nagoya Protocol to the CBD, for example, contains rules on mutually agreed terms and the practicalities of benefit-sharing contracts. Guidelines about consultation and consent remain in separate instruments, such as the Akwé: Kon Guidelines (Convention on Biological Diversity, 2004). Legal scholars may of course argue otherwise on the basis of their specialized knowledge, but the image for the non-specialist remains fragmented, and the idea of benefit-sharing as a process is not conveyed. Arguments about the need to read benefit-sharing in tandem with human rights instruments emerge as particularly important in this view (Morgera, 2018).

The aim of this chapter was to present the common macro-level themes that emerged from a comparison of case studies concerning very different communities, dealing with very different scenarios linked to benefit-sharing. The breadth of the cases studied suggests that these common themes may have some more general applicability, though this cannot be proven on the basis of five cases alone. Nevertheless, many of the common themes are echoed in other literature examining local case studies of benefit-sharing (Wynberg and Hauck, 2014; Robinson, 2015; Tysiachniouk and Olimpieva, 2019). Whether they are generally applicable or not, this suggests it is worth considering these common themes when examining the international level of environmental governance with a view to asking whether it hosts spaces for issues important to local communities. Given the breadth of the themes that emerged from the case comparison, the consideration of the international level presented in the following two chapters follows this lead. Local communities underline a range of issues linked to benefit-sharing, and a range of areas at the international level are thus relevant. The next chapter presents an evaluation of where spaces for issues considered important by local communities that emerged from this case comparison exist in the CBD – the global environmental governance arena that is arguably most open to local voices – by way of beginning this task.

Notes

1 This discussion draws on Marchegiani, Morgera and Parks (2019).
2 National Constitution of Argentina, art. 75, point 17: 'To recognize the ethnic and cultural pre existence of indigenous peoples of Argentina.'
3 See the organization's mission at https://ikariandocs.wordpress.com/about-ikarian-documentationresearch-and-action-center/ (accessed 15 July 2019).

4 The argument for support from State actors is impossible to elaborate upon on the basis of these cases, since only one State actor was providing support in the Namibian context.
5 https://www.boell.de/en (accessed 20 May 2019).
6 See Marazuela *et al.*, 2018 – lithium-mining techniques use large amounts of water, and have the potential to cause mixing between fresh and saltwater.
7 Environmental concerns were expressed in the country's development goals in terms of poverty reduction and food security at that time.

References

Anaya, J. (2010) *Report on the situation of human rights and fundamental freedoms of indigenous people.* UN Doc. A/HRC/15/37/.

Anaya, J. (2012) *Report of the Special Rapporteur on the rights of indigenous peoples, Progress report on extractive industries*. UN Doc. A/HRC/21/47.

Bavikatte, K., Robinson, D. F. and Oliva, M. J. (2015) 'Biocultural community protocols: dialogues on the space within', *IK: Other Ways of Knowing,* 1(2), pp. 1–31.

Bayrak, M. M., Tu, T. N. and Burgers, P. (2013) 'Restructuring space in the name of development: the sociocultural impact of the Forest Land Allocation Program on the indigenous Co Tu people in Central Vietnam', *Journal of Political Ecology*, 20(1), pp. 37–52.

Bosi, L., Giugni, M. and Uba, K. (eds) (2016) *The Consequences of Social Movements.* Cambridge, UK: Cambridge University Press.

Choudry, A. and Kapoor, D. (2013) *NGOization: Complicity, Contradictions and Prospects.* London: Zed Books.

Cocchiaro, G. *et al.* (2014) 'Consideration of a Legal Trust Model for the Kukula Healers' TK Commons in South Africa', in de Beer, J. *et al.* (eds) *Innovation and Intellectual Property: Collaborative Dynamics in Africa.* Cape Town: University of Cape Town Press, pp. 151–170.

Comunidades Indígenas de las Salinas Grandes y Laguna de Guayatayoc de Jujuy y Salta (2015) *Kachi Yupi/Huellas de la Sal: Procedimiento de Consulta y Consentimiento Previo, Libre e Informado para las Comunidadades Indígenas de la Cuenca de Salinas Grandes y Laguna de Guayatayoc.* Available at: https://cl.boell.org/sites/default/files/protocolo_final.pdf.

Convention on Biological Diversity (2004) *Akwé: Kon Voluntary Guidelines for the Conduct of Cultural, Environmental and Social Impact Assessment Regarding Developments Proposed to Take Place on, or which are Likely to Impact on, Sacred Sites and on Lands and Waters traditionally Occupied or Used by Indigenous and Local Communities.* CBD Decision VII/16, Annex, Montreal.

Convention on Biological Diversity (2010) *The Nagoya Protocol on Access to Genetic Resources and the Fair and Equitable Sharing of Benefits Arising from their Utilization to the Convention on Biological Diversity.* CBD Decision X/1, Annex, Montreal.

Convention on Biological Diversity (2016) *Mo'otz Kuxtal Voluntary guidelines for the development of mechanisms, legislation or other appropriate initiatives to ensure the "prior and informed consent", "free, prior and informed consent" or "approval and involvement", depending on national circumstances, of indigenous peoples and local communities for accessing their knowledge, innovations and practices, for fair and equitable sharing of benefits arising from the use of their knowledge, innovations and practices relevant for the conservation and sustainable use of biological diversity, and for reporting and preventing unlawful appropriation of traditional knowledge*. CBD Decision XIII/18, Annex, Montreal.

Gilbert, J. and Doyle, C. (2011) 'A New Dawn over the Land: Shedding Light on Collective Ownership and Consent', in Allen, S. and Xanthaki, A. (eds) *Reflections on the UN Declaration on the Rights of Indigenous Peoples*. Oxford: Hart, pp. 289–328.

International Labour Organization (1989) *Convention no. 169 Concerning Indigenous and Tribal Peoples in Independent Countries*. 28 ILM 1382.

Jonas, H., Bavikatte, K. and Shrumm, H. (2010) 'Community protocols and access and benefit sharing', *Asian Biotechnology and Development Review*, 12(3), pp. 49–77.

Keck, M. E. and Sikkink, K. (1998) *Advocacy Networks in International Politics*. Ithaca, NY: Cornell University Press.

Marazuela, M. A. *et al.* (2018) 'Hydrodynamics of salt flat basins: the Salar de Atacama example', *Science of the Total Environment*, 651(1), pp. 668–683.

Marchegiani, P., Hoglund Hellgren, J. and Gomez, L. (2018) *Lithium extraction in Argentina: a case study on its social and environmental impacts*. Available at: https://goodelectronics.org/wp-content/uploads/sites/3/2019/05/DOC_LITHIUM_ENGLISH.pdf.

Marchegiani, P., Morgera, E. and Parks, L. (2019) *Indigenous peoples' rights to natural resources in Argentina, the challenges of impact assessment, consent and fair and equitable benefit-sharing in cases of lithium mining*, BENELEX Working Papers, 19. Available at: https://papers.ssrn.com/sol3/papers.cfm?abstract_id=3317375.

Meyer, D. S. (2004) 'Protest and political opportunities', *Annual Review of Sociology*, 30, pp. 124–145.

Morgera, E. (2016) 'The need for an international concept of fair and equitable benefit-sharing', *European Journal of International Law*, 27(2), pp. 353–383.

Morgera, E. (2017) *Reflections on 2016 UN Biodiversity Conference (Parts I & II): assessing the Mo'otz kuxtal guidelines on benefit-sharing from the use of traditional knowledge*, BENELEX Blog.

Morgera, E. (2018) 'Dawn of a new day? The evolving relationship between the Convention on Biological Diversity and international human rights law', *Wake Forest Law Review*, 54, pp. 101–121.

Murphree, M. (2004) 'Communal approaches to natural resource management in Africa: from whence to where?', *Journal of International Wildlife Law and Policy*, 7, pp. 203–216.

Nelson, F. (2010) 'Democratizing Natural Resource Governance: Searching for Institutional Change', in Nelson, F. (ed.) *Community Rights, Conservation and Contested Land: The Politics of Natural Resource Governance in Africa*. London: Earthscan, pp. 310–333.

Ostrom, E. (2010) 'Beyond markets and states: polycentric governance of complex economic systems', *American Economic Review*, 100(3), pp. 641–672.

Parks, L. and Morgera, E. (2015) 'The need for an interdisciplinary approach to norm diffusion: the case of fair and equitable benefit-sharing', *Review of European, Comparative and International Environmental Law*, 24(3), pp. 353–367.

Ribot, J. C. and Peluso, N. L. (2003) 'A theory of access', *Rural Sociology,* 68(2), pp. 153–181.

Robinson, D. F. (2015) *Biodiversity, Access and Benefit-Sharing. Global Case Studies*. London: Routledge.

Swemmer, L., Grant, R., Annecke, W., and Freitag-Ronaldson, S. (2015) 'Toward more effective benefit-sharing in South African national parks', *Society & Natural Resources,* 28(1), pp. 4–20.

Tsioumani, E. (2020) *Fair and Equitable Benefit-sharing in Agriculture: Reinventing Agrarian Justice*. London: Routledge.

Tysiachniouk, M. and Olimpieva, I. (2019) 'Caught between traditional ways of life and economic development: interactions between indigenous peoples and an oil company in Numto Nature Park', *Arctic Review on Law and Politics,* 10, pp. 56–78.

Vermeylen, S. (2007) 'Contextualizing "fair" and "equitable": the San's reflections on the Hoodia benefit-sharing agreement', *Local Environment,* 12(4), pp. 423–436.

Vermeylen, S. (2017) 'Materiality and the Ontological Turn in the Anthropocene: Establishing a Dialogue between Law, Anthropology and Eco-philosophy', in Kotzé, L. (ed.) *Law and Governance for the Anthropocene*. Oxford: Hart.

Wynberg, R. and Hauck, M. (eds) (2014) *Sharing Benefits from the Coast. Rights, Resources and Livelihoods*. Cape Town: University of Cape Town Press.

Young, I. M. (2004) *Inclusion and Democracy*. Oxford: Oxford University Press.

4 Local voices, international arenas

The case of the Convention on Biological Diversity

This chapter moves the focus of the book to the international level, and more precisely to the CBD. Five local case studies were presented in Chapter 2, and common themes arising from them were presented in Chapter 3. These common themes, it was argued, are important to the local communities studied and, perhaps, beyond. This chapter reflects on the existence and extent of discursive spaces for these common themes in the CBD. It presents the results of a detailed discourse analysis that draws on the literature reviewed in Chapter 1.[1] The aim of the analysis is to draw a discursive 'map' of the CBD over time. This indicates areas where the CBD reflects the view of the critical literature, and areas where it challenges it.

In brief, the critical literature (see Chapter 1: Local communities and global environmental governance) argues that certain discourses shape and delimit the boundaries of global decision-making on environmental issues. The dominant discourses identified are rooted in a view of the earth or the environment as separate from society or culture, and a market or capitalist worldview. Some scholars argue that this applies at least in part to the CBD (e.g. Brand and Vadrot, 2013; Marion Suiseeya, 2014). Others, however, argue that the CBD is not dominated by these discourses to the same extent as other international bodies, and is thus more accessible for local themes and voices. This is argued on the basis of the text of the treaty itself (Reimerson, 2013), the adoption of the Nagoya Protocol (e.g. Jonas, Bavikatte and Shrumm, 2010; Bavikatte, 2014), and the CBD's relative openness to non-state actors (Affolder, 2017). On this last point, Schabus argues that the CBD provides an 'unprecedented space for indigenous participation' in particular (2017: 26).

The discourse analysis of the CBD presented in this chapter reveals a detailed picture of where and when the CBD conforms more closely with the critical view of global environmental governance, and where it does not. The assumption is that in those areas where the CBD is more discursively open, there will be more space for raising the common themes important to local communities, discussed in Chapter 3. The chapter reflects on how some of these common themes could be introduced at the international level of the CBD within these discursive spaces. Not all of the common themes are addressed in equal depth: governance contexts and support from non-state actors, for example, are also covered in the reflections in Chapter 5 (The view from above – participation and local communities in the CBD) on paths to participation. Here, those common themes that are more outward-facing, in the sense that they require action by other actors as well as local communities, are discussed in greater detail: recognition from without, procedural guarantees, non-monetary benefits, and the view of benefit-sharing as a process.

The CBD is the chosen focus for the analysis precisely because there is some divergence in the literature about the spaces it provides to local communities and indigenous peoples. For this reason, it is an interesting starting point to answer questions about the existence of spaces where local issues can be raised, and eventually addressed, through global policies such as benefit-sharing. Signed in 1992 and in force in December 1993, the CBD is one of the so-called Rio treaties alongside those dealing with climate change and desertification. There are 196 parties to the CBD, and it is the closest to universal membership of the three Rio treaties.[2] The CBD is a framework treaty – it sets general objectives and guiding principles. These are developed in successive instruments such as protocols and COP decisions. Although the CBD is a legally binding treaty, it has been described as following a soft approach (Harrop and Pritchard, 2011). This is also explained with reference to its broad scope, which led to 'many ambiguities and omissions, and highly qualified commitments' in its text (Parks and Schröder, 2018: 750; see also Birnie, Boyle and Redgewell, 2009).

One result of the wide scope and soft approach of the CBD is the view that the CBD leaves spaces for different stakeholders to participate in its work. The three objectives of the CBD, stated in Article 1, are 'the conservation of biological diversity, the sustainable use of its components and the fair and equitable sharing of the benefits arising out of the utilization of genetic resources'. Benefit-sharing is also to be pursued through technology transfers and funding. As far as local communities are concerned, Article 8(j) of the CBD refers to the need to 'respect, preserve and maintain the knowledge, innovations and practices of indigenous and local communities' where these are useful for conservation. It also underlines that parties should promote the wider application of these, with the 'approval and involvement' of knowledge holders. However, the text is qualified in that all of this should be done only where in line with national legislation. This is problematic given that many countries do not recognize indigenous peoples – a point raised by scholars who argue the CBD leaves little space for local communities in real terms (see Chapter 1: Local communities and global environmental governance). Over time, COP decisions have added detail about the implementation of Article 8(j) and its related provisions. Notably, the COP has adopted the Nagoya Protocol on Access and Benefit-sharing (Convention on Biological Diversity, 2010a), as well as instruments including: voluntary guidelines on cultural, environmental and social impact assessments (the Akwé: Kon Guidelines: Convention on Biological Diversity, 2004b); principles and guidelines on the sustainable use of biodiversity (the Addis Ababa Guidelines: Convention on Biological Diversity, 2004a); a code of ethical conduct (the Tkarihwaié:ri Code: Convention on Biological Diversity, 2010b); and guidelines on prior and informed consent (the Mo'otz Kuxtal Guidelines: Convention on Biological Diversity, 2016). At its most recent meeting, the COP adopted guidelines on the repatriation of traditional knowledge (the Rutzolijirisaxik Guidelines: Convention on Biological Diversity, 2018).

As the most developed international source of law on benefit-sharing and related issues, the CBD is also an ideal candidate for the present analysis to the extent that it acts as a frontrunner for other arenas of global environmental governance. If the CBD contains spaces where local concerns can be raised, it may act as a model of best practice for other multilateral environmental agreements where questions of benefit-sharing and other issues of broad importance for local communities are concerned. This would be in line with the argument that the CBD has acted in this manner for the involvement of non-state actors. Through bodies such as the Ad Hoc Working Group on Article 8(j) on traditional knowledge, innovations and practices, where representatives of indigenous peoples and local communities are active participants and contribute to the development of instruments of direct relevance to them, the CBD has been said to set a precedent for participation (e.g. Affolder, 2017).

It is with these arguments in mind that the discourse analysis of the CBD is carried out. This is not to claim that discourse alone will lead to direct impacts for how these themes are raised or eventually resolved. Rather, the presence or absence of discursive grounds for open discussions of these themes is seen as crucial to eventual responses. The paths through which local voices might be heard in the CBD are discussed in Chapter 5, which looks at the sites, types, and forms of participation the CBD envisages for local voices, along with the potential use of community protocols as conduits for local voices. The following sections present the method used to carry out the discourse analysis before presenting results and reflections on what these mean for the common themes emerging from the case study comparison described in Chapter 3.

Analyzing the discourses of the Convention on Biological Diversity

Before looking in detail at this discursive map of the CBD, this section provides more information on the method and underlying methodology used, as introduced in Chapter 1 (Analyzing space for local themes and participation at the international level). Discourse analysis has been used to map environmental politics, policy and law by a number of scholars

(Dryzek, 2005; Hajer and Versteeg, 2005; Gellers, 2015). The method of discourse analysis is not a standardized one with a single, agreed rulebook (Moses and Knutsen, 2012). Rather, it is designed on an ad hoc basis to fit particular research data and questions, and can be more or less formal, comprising a range of approaches. This elasticity is related to its ontological roots in constructivism. Theories of discourse, as already discussed in Chapter 1 (Analyzing space for local themes and participation at the international level), posit that these shape and constrain our decisions, but can be challenged by minority discourses and thus change over time. Discourses are not necessarily explicit, but may be implicit in a text or other source. They can thus be identified using different analytical approaches, but all include a strong element of researcher subjectivity. To address this, discourse analyses should follow clear and transparent methods. The description of the method used here thus seeks to reduce subjectivity in this way.[3]

The discourse analysis presented here is applied to CBD COP decisions. The COP decisions are sampled, so that only those parts of the text relevant for indigenous peoples and local communities (to use the term currently adopted by the CBD) are analyzed. COP decisions are an ideal data source for two main reasons: first, because much of the existing work discussing the discourses that underpin the CBD focuses on the text of the treaty itself; second, as already mentioned in the introduction to this chapter, because COP decisions give shape to the text of the treaty, and detail how it is to be implemented. They are thus a good source for looking at the dynamic ways in which discourses change over time in the context of the CBD regime (Morgera and Tsioumani, 2011; Parks, 2018). Finally, COP decisions are the result of long discussions in various working groups and other bodies that culminate in consensus decision-making by the parties to the CBD – they represent the final decisions prevailing on specific matters at the time of their adoption, and thus indicate which discourses characterized that prevailing opinion.

Each of the texts sampled from the COP decisions was 'coded' according to whether they were deemed to point to one or more of the elements of the discourse argued to characterize the CBD in the critical literature. These elements, drawn from the critical literature, are each accompanied by elements drawn in turn from the more optimistic literature to form 'pairs' of discourse categories (see Parks, 2018, for further detail). The pairs of discourse categories are: exclusion versus participation; external imposition versus internal initiatives; modern or western science versus recognition of traditional knowledge; and finally capitalist or market reasoning versus valorization of other worldviews (ibid.).

The exclusion category (and its counterpart – participation) draws from literature that highlights the ways in which discourses define the voices that are suitable to be heard in decision-making processes. Much of this scholarship points to the colonial period as the origin of a hegemonic discourse of exclusion via the 'othering' of indigenous peoples and local communities.

Uggla (2010), for example, highlights the discursive creation of a nature/culture divide where 'untouched' nature is enclosed for its protection (an approach also known as 'fortress conservation'). This fails to attribute agency to local residents in the shaping and protection of lands and waters, and is argued to continue to characterize international environmental decision-making arenas in the present. The exclusion category was assigned to texts that name other actors as part of decision-making processes but fail to mention indigenous peoples and local communities.

There is some bias in the sample here, however, as the texts were selected on the basis of their relevance to indigenous peoples and local communities – their exclusion from other COP decisions may occur elsewhere, and the results should be read with this borne in mind. The analysis does take note, however, of other pieces of information that provide further clues about potential exclusion. The CBD has defined a total of 24 cross-cutting issues and 7 different thematic programmes on major biomes, and COP decisions are mapped to these.[4] One of these, 'traditional knowledge, innovations and practices – Article 8(j)' (from now on traditional knowledge), is the natural home of decisions that are relevant for indigenous peoples and local communities, since Article 8(j) is the basis for consideration of indigenous peoples and local communities in the CBD. Reflecting on the extent to which discourses about the exclusion and participation of local communities correspond with different cross-cutting issues provides clues about whether they are 'ghettoized' or confined to a particular area of the CBD. The discourse categories of exclusion and participation speak to a number of the common themes that emerged from the case comparison. Both are relevant to aspects discussed under the heading of recognition in Chapter 3. They are also linked to procedural guarantees, since exclusion or participation has effects on access to information and clear consent procedures in national contexts and, in turn, on the view of benefit-sharing as a process.

The category of external imposition and its counterpart 'internal initiative' build on the exclusion and participation categories. Indeed, local communities can be excluded not only from decision-making processes, but also from the implementation of the resulting decisions. External imposition draws on discussions in the literature on CBNRM that highlight the ineffectiveness of policies imposed by external actors with little knowledge of local realities, particularly when this occurs in the absence of effective transfers of power and real, substantively democratic, involvement (Murphree, 2004; Nelson, 2010; see Chapter 1: Local communities and global environmental governance). Although the literature on CBNRM focuses on national policies, similar risks have been noted for policies stemming from the international level, and the CBD in particular, including benefit sharing (Adger *et al.*, 2001). External imposition can be noted in COP decisions that call on local communities to play an active role in implementing policies formed without their involvement. The 'internal initiative' category, on the other hand, draws on work taking a more optimistic view of the role of local communities in the CBD and its application. Scholarship in this vein has underlined areas where local communities' approaches to conservation, based on detailed local and traditional knowledge, are encouraged. 'Internal initiatives' of this kind include community protocols, recognized in the Nagoya Protocol (Bavikatte, 2014), territories and areas conserved by indigenous peoples and local communities (ICCAs) (Jonas, 2017), and others. These categories also speak to issues around external recognition and procedural guarantees, particularly as they relate to recognition from external actors. They also tie in with the question regarding organization and the roles of external actors. Internal initiative discourses in particular may provide a basis for discussions of long-term funding for community organizations, as well as about the role of external actors in supporting local communities, while external imposition may have the opposite effects. As before, these questions also feed into the view of benefit-sharing as a process.

The categories of modern or western science and recognition of traditional knowledge and customary sustainable use form the next pair. Critical literature on environmental governance links the exclusion of local communities to a nature/culture divide. In this view, nature is subject to human exploitation and manipulation through western models of science rooted in Enlightenment experimental methods. This preference for technical, western scientific solutions is seen to characterize international environmental decision-making in the present. Marion Suiseeya (2014), for example, notes that discussions of the Nagoya Protocol were dominated by technical debates, while Brand and Vadrot (2013) make a similar point about the exclusion of traditional knowledge. The recognition of traditional knowledge, on the other hand, would create a discursive basis for challenging or redefining western scientific approaches. This raises an important point about the discourse analysis method – that categories can be combined. Thus, the recognition of traditional knowledge can be assigned in combination with modern or western science discourses to signal that opening up to a plurality of knowledge sources may be possible. These categories speak to a different aspect of the central recognition issue that emerged from the case comparison – that of respect. They can also be linked to elements of procedural guarantees, particularly consent procedures geared to genuine dialogue and understanding.

The link to recognition as respect is also a clear consideration for the categories of capitalist or market reasoning and the valorization of other worldviews. These categories try to capture the claim made in the critical literature that capitalism, also expressed via colonialism, is the dominant worldview underpinning global environmental governance. Evidence that this is the case is pointed to in green economy approaches within the wider sustainable development discourse. The latter conceives of environmental protection and social justice goals as possible without any sacrifice of continuous economic growth, so long as environmental goods are assigned their appropriate (economic) value and treated as precious commodities (Dryzek, 2005; UNEP, 2011). According to more critical views, such a 'win–win' situation is unrealistic, as many critical environmental problems faced today originated in the capitalist era. This means that hopes for a capitalist approach to resolve environmental problems are false at best, and perpetuate dominant global North–South relations at worst (Natarajan and Khoday, 2014). Policies such as payments for ecosystems services and indeed access and benefit-sharing are interpreted by some scholars in this light (e.g. Svarstad, 2004; Wilkinson, 2014), though others underline that they do comprise more holistic conceptions of complex environmental problems in an otherwise fragmented approach (Morgera and Savaresi, 2013). Identifying texts that valorize 'other worldviews' is challenging, since by definition these other worldviews are many and varied. The category was therefore assigned to texts that conveyed a sense of an understanding of nature and culture as mutually dependent and indivisible, in line with the critical literature (e.g. Uggla, 2010). Information about where the CBD signals capitalist or market reasoning, and where it signals recognition of other worldviews beyond this, is useful for evaluating spaces for some of the common themes presented in Chapter 3. In particular, it provides information about recognition as respect for other worldviews or cosmologies, and the importance of non-monetary benefits.

The following sections illustrate the results of the discourse analysis of CBD COP decisions. The section begins with an overview of the results, followed by a discussion focusing on the discourse categories drawn from the more optimistic literature, then the critical literature. What the results indicate in terms of discursive spaces for the common themes that emerged from the case comparison is addressed throughout. A final section presents some preliminary findings based on an analysis comprising all of the texts of the COP decisions to contextualize these results.

Reflections on a discursive map of the Convention on Biological Diversity

Figure 4.1 shows the extent to which each of the discourse categories presented in the previous section characterizes texts relevant to indigenous peoples and local communities in CBD COP decisions over time, stretching from the first COP in 1994 to the fourteenth COP in 2018. At first glance, Figure 4.1 suggests a rather robust challenge to the critical literature given the clear predominance of the participation and recognition categories, yet closer inspection combined with some qualitative analysis reveals a more nuanced picture. The following sections discuss this in detail.

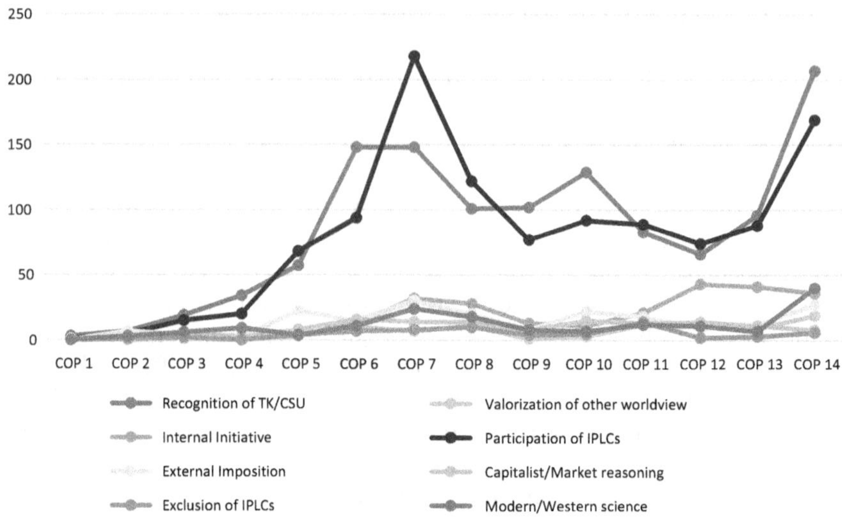

Figure 4.1 An overview of the discourse analysis for all categories.

Discourse categories that challenge the critical literature

What emerges most clearly from Figure 4.1 is that when COP decisions talk about indigenous peoples and local communities, they tend to focus on themes of recognition (of traditional knowledge and customary sustainable use – TK/CSU) and participation (Parks, 2018). Although there are large fluctuations in their frequencies, recognition and participation are the most common discourse categories over time, and whenever they are recorded they 'stick' in the sense that COP decisions build on one another rather than overturning previous decisions. Where decisions are of use to indigenous peoples and local communities, they thus remain so over time – though the same applies to decisions that threaten spaces for raising local issues. Beginning with recognition, the link to the central theme of the local-level case studies is clear: the CBD recognizes the role of traditional knowledge and customary sustainable use in the conservation of biodiversity. This is the essence of Article 8(j), and is an affirmation recalled throughout COP decisions. Three notable high points can be seen in Figure 4.1, corresponding with moments when COP decisions refer to issues of direct relevance to matters of recognition: an outline of the Report on the Status and Trends Regarding the knowledge, innovations and practices of indigenous and local communities relevant to the conservation and Sustainable Use of Biodiversity (Convention on Biological Diversity, 2002) and the Bonn Guidelines on access to genetic resources and fair and equitable sharing of the benefits arising out of their utilization (UNEP, 2002) in COP 6; the Akwé: Kon Guidelines and the Addis Ababa Principles (Convention on Biological

Diversity, 2004b, 2004a) in COP 7; and the Rutzolijirisaxik Voluntary Guidelines for the Repatriation of Traditional Knowledge in COP 14 (Convention on Biological Diversity, 2018).

There is thus a relatively stable level of recognition for traditional knowledge and customary sustainable use from COP 5 (2002) on, with a notable surge in COP 14 (2018) as the Strategic Plan for Biodiversity 2011–2020 comes to a close. The recognition in the CBD COP is 'from without', to use the language of Chapter 3 – it is recognition about the contribution made by local communities, but only insofar as this is based on traditional knowledge and customary sustainable use. It is a limited form of recognition, both in terms of the specific basis of local community contributions, and in the mention of the need to stay in line with national legislation included in Article 8(j). Nevertheless, this is still recognition 'from without' and could conceivably be referred to in all of the case studies presented, all of which involve communities which use traditional knowledge or approaches to conservation. The recognition of local communities in the CBD is used in some community protocols to underline that the positions of local communities are bolstered by international law. This can strengthen claims for recognition from external actors closer to home. The CBD is mentioned, for example, in the original community protocol of the Kukula Association (Traditional Health Practitioners of Bushbuckridge, 2010), though the South African legislative backdrop is also conducive to community recognition (see Chapter 3: Complex governance contexts). The CBD has also been noted as a useful springboard for community claims in areas where binding protocols exist, for example on the Nagoya Protocol on access and benefit-sharing from genetic resources (Robinson, 2015). While there is clearly discursive space for the common theme of recognition to be raised at the international level, recognition is clearly delineated and constrained. When invoking international recognition to support local claims, the same delineations could be applied.

The second discourse category that characterizes the COP decisions is participation. Participation has general relevance for the common themes that arose from the comparison of the case studies. If local communities can participate in international decision-making, they have the potential ability to raise all kinds of issues, even where these fall outside the normal discursive boundaries of the arena in question, and can thus challenge dominant discourses. Participation is also important for recognition from external actors. Here, participation could be argued to equate with recognition of local communities beyond their value for conservation through traditional knowledge and customary sustainable use. If local actors are seen as valid participants in international processes, it could be claimed that recognition has become more generalized, constructing local communities as stakeholders in a wider sense.

While the prevalence of texts referring to the participation of indigenous peoples and local communities suggests that a view of these groups as stakeholders in a more general sense has taken root, a closer look at their correspondence with cross-cutting issues tempers this. As mentioned in the introduction to this chapter, CBD COP decisions refer to a number of cross-cutting issues, and texts about indigenous peoples and local communities originated in the traditional knowledge cross-cutting issue. One source of information to gauge the extent to which COP decisions see indigenous peoples and local communities as general stakeholders in the CBD's work is the correspondence of texts assigned the participation category with cross-cutting issues beyond traditional knowledge. Previous analysis of COP decisions up to the thirteenth meeting in 2016 has shown that texts assigned the participation category are 'ghettoized' (Parks, 2018). While traditional knowledge is itself a cross-cutting issue, the presence of a participation discourse in other issue areas would nevertheless indicate a more widely held view of indigenous peoples and local communities as stakeholders. To evaluate this including data from COP 14, Figures 4.2 and 4.3 display the distribution of texts mentioning participation for local communities amongst cross-cutting issues at the two peaks of COP 7 and COP 14. These high points of talk about local community participation give an idea of the extent of spread of this discourse category at these moments.

Figure 4.2 shows that texts referring to local community participation are mostly confined to the traditional knowledge cross-cutting issue in COP 7, in line with previous findings (Parks, 2018). Over half of the texts mentioning the participation of indigenous peoples and local communities adopted at this meeting fall into the traditional knowledge cross-cutting issue, with significantly lower numbers appearing in 11 other areas. In COP 14, the majority of texts mentioning participation still fall within the

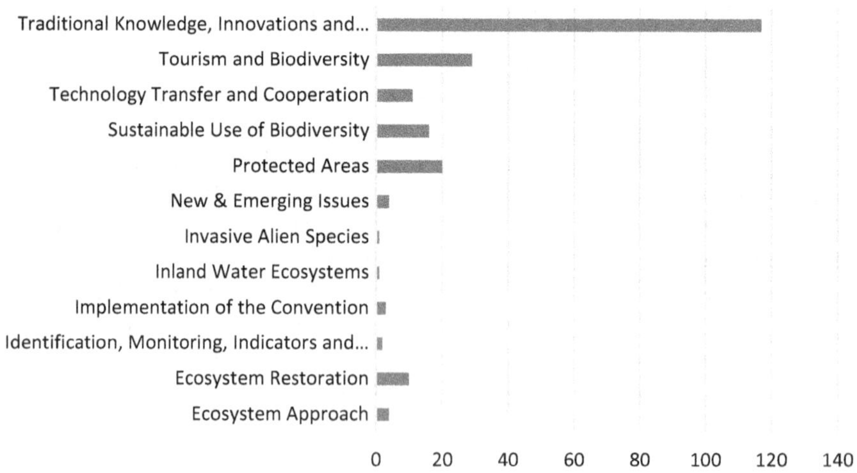

Figure 4.2 Participation - distribution across cross-cutting issues in COP 7.

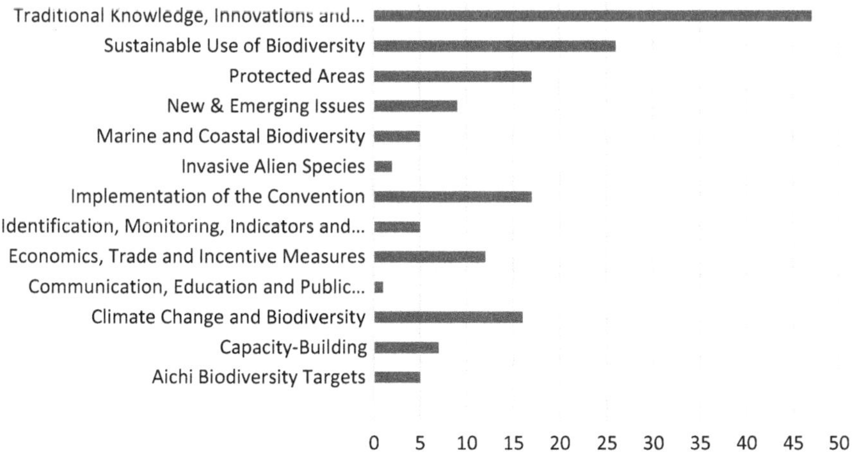

Figure 4.3 Participation – distribution across cross-cutting issues in COP 14.

traditional knowledge cross-cutting issue, but these now account for fewer than a third of the total. Distribution is far from even, but certainly represents a more varied spread compared to COP 7. The beginnings of a movement towards the cross-cutting issues of sustainable use of biodiversity and protected areas, visible in the data on COP 7 in Figure 4.2, thus appear to be built upon in COP 14. Other cross-cutting issues where participation is mentioned are new in COP 14 compared to COP 7, such as climate change and biodiversity. Still others have a stronger showing compared to COP 7, for example in the cross-cutting issue implementation of the Convention. Some of these differences are attributable to the agendas set for each COP, since not all meetings cover all issues in equal depth. Yet the cross-cutting issues covered do relate to all three of the CBD's objectives: traditional knowledge, sustainable use, and protected areas are linked to the conservation of biological diversity and benefit-sharing, economics, trade, and incentives measures also speak to benefit-sharing and technology transfer, and new and emerging issues are also covered. A general trend thus appears to emerge, suggesting that the idea of participation for indigenous peoples and local communities is spreading to different areas of the CBD over time. Indigenous peoples and local communities are thus increasingly considered as stakeholders. Although this is primarily rooted in Article 8(j), and thus on the basis of their traditional knowledge and practices which contribute to conserving biodiversity, local community participation on the basis of the roles they play in other areas of work beyond Article 8(j) is attributed to a smaller but increasing extent.

If local communities are coming to be considered as stakeholders beyond their roles as holders of traditional knowledge, they will have a broader basis to raise common themes such as procedural guarantees, even where these are linked to other roles - for example those described in the case studies in this book. The relevance of the participation category for procedural guarantees is also underlined by the distribution of texts illustrated in Figure 4.1. Texts mentioning participation have two clear peaks. The first is at COP 7, and coincides with (amongst other things) the Akwé: Kon Guidelines, whose full title refers to the Conduct of Cultural, Environmental and Social Impact Assessment Regarding Developments Proposed to Take Place on, or which are Likely to Impact on, Sacred Sites and on Lands and Waters traditionally Occupied or Used by Indigenous and Local Communities (Convention on Biological Diversity, 2004b). The guidelines outline best practices for impact assessments, including the procedural guarantees needed in these processes discussed in Chapter 3. The second peak is at COP 14, when the Rutzolijirisaxik Voluntary Guidelines for the Repatriation of Traditional Knowledge were adopted (Convention on Biological Diversity, 2018). Here too, procedural guarantees form a central element of best practices.

The strong showings of the recognition and participation discourse categories over time suggest that some discursive spaces are available for local communities to raise some of the common themes that emerged from the case comparison, and indeed for local communities to take part in the CBD. Questions remain about how broad these spaces might be beyond the role of local communities as traditional knowledge holders, as well as about the actual paths for participation envisaged by the CBD (the latter is discussed in more depth in Chapter 5: The view from above). Two other discourse categories provide further clues about these, namely those of 'valorization of other worldview' and 'internal initiative'. Their respective trajectories are illustrated in Figure 4.4.

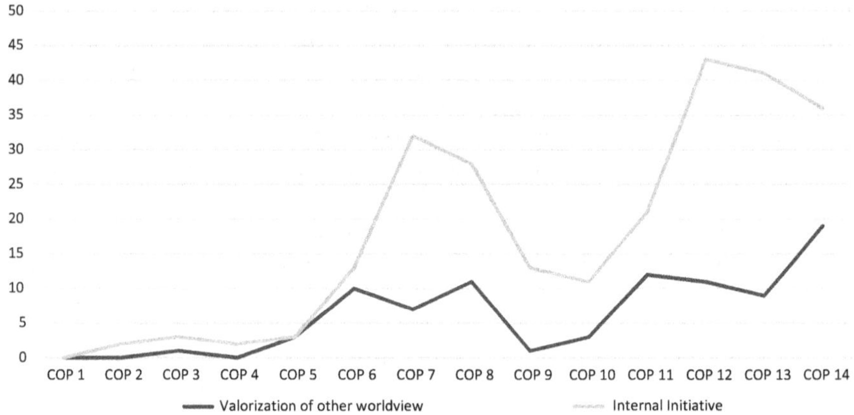

Figure 4.4 Overview of discourse categories 'valorization of other worldview' and 'internal initiative'.

Texts assigned the category of 'valorization of other worldview' provide more insight on the question of how circumscribed discursive spaces are for locally relevant matters, such as the common themes explored in Chapter 3. If the COP decisions were found to recognize different worldviews as relevant for the treaty's aims, this would suggest spaces to raise a wider range of themes. In other words, if local communities' worldviews (where they differ from the dominant, market-based view argued to underpin the CBD and other arenas of global environmental governance) are respected in the CBD's COP decisions, their definitions of and proposed solutions to environmental problems may equally be respected and considered. This was a key aspect in the central theme of recognition that emerged from the local case studies. The recognition of different worldviews also speaks to other themes that emerged from the comparison. If different outlooks find discursive spaces, so may diverse modes of organizing and decision-making, and as a result different needs about procedural guarantees including timings, information formats, and other aspects. In addition, this discourse category is also linked to the importance of non-monetary benefits and the view of benefit-sharing as a process. The importance attached to a range of benefits by the local communities discussed in Chapter 3 is closely related to worldviews and definitions of 'good' or 'benefit'. Respect for other worldviews should thus allow discursive spaces for definitions of benefit beyond the monetary.

As illustrated in Figure 4.4 (and alongside all the other discourse categories in Figure 4.1), texts that recognize the validity of non-dominant worldviews are not common in CBD COP decisions. As discussed in the introduction to this chapter, only those texts that conveyed a sense of an understanding of nature and culture as mutually dependent and indivisible were assigned this category. This may go some way to explaining the low numbers, since such philosophical statements are arguably unlikely to appear in legal texts like COP decisions. Nevertheless, the general trend is an increasing one, albeit in fits and starts. As the numbers are low, a more detailed overview of the content of the texts concerned is possible, and provides information about a qualitative change.[5]

The first texts assigned this category, beginning in COP 3, refer to 'market and non-market values of biological diversity' and how these should be integrated into the CBD's work. Other worldviews are thus defined with reference to the market, a trend that continues until COP 6.[6] From this point, texts that move away from market terminology begin to take precedence. In COP 8, texts demonstrating an understanding of a link between the protection of biodiversity and the worldviews underpinning the social systems that produce and achieve this appear. For example, one text refers to islands as 'home to unique cultures that have developed traditional resource-management methods that have, in many cases, enabled people to develop and live in harmony with biodiversity'. This idea is developed further in COP 10, in the Tkarihwaié:ri Code of Ethical Conduct

(Convention on Biological Diversity, 2010). The Code includes statements such as the following:

> [t]raditional guardianship/custodianship recognizes the holistic interconnectedness of humanity with ecosystems and obligations and responsibilities of indigenous and local communities, to preserve and maintain their traditional role as traditional guardians and custodians of these ecosystems through the maintenance of their cultures, spiritual beliefs and customary practices.

The Nagoya Protocol, adopted at the same meeting, also includes language underlining 'the interrelationship between genetic resources and traditional knowledge' and 'their inseparable nature'.

Moving forward to COP 13, texts calling for respect for other cosmologies appear, along with references to a dialogue held at the CBD Ad Hoc Open-ended Working Group on Article 8(j) entitled 'challenges and opportunities for international and regional cooperation in the protection of shared traditional knowledge across borders for the strengthening of traditional knowledge and the fulfilment of the three objectives of the Convention, in harmony with Nature/Mother Earth'. This line of reasoning appears again in COP 14, though many of the texts assigned this category at this meeting are more matter of fact about taking 'worldviews' into account. For example, one text notes that

> Indigenous peoples and local communities have managed variability, uncertainty and change through multigenerational histories of interaction with the environment [...] Integrating the knowledge of indigenous peoples and local communities also involves an appreciation of their *cosmovisión* (A worldview that has evolved over time that integrates physical and spiritual aspects).

While the frequency of texts that recognize and validate other worldviews remains low, there is thus some evidence of a positive qualitative change, as texts progress from defining other worldviews in opposition to market values to more meaningful and in-depth descriptions.

The second discourse category illustrated in Figure 4.4 is 'internal initiative'. This category is, as explained earlier in the chapter, assigned to texts that move beyond the idea of the participation of indigenous peoples and local communities in the CBD's work to encourage and support their own initiatives that contribute to its aims. This sort of bottom-up initiative, developed by a local community on its own terms, includes community protocols and various others. Texts assigned the 'internal initiative' category can provide information about how the CBD recognizes and encourages indigenous peoples and local communities' efforts to solve environmental issues in locally tailored ways. In recognizing, supporting, and encouraging internal initiatives, the CBD would address several of the common themes discussed in Chapter 3. Recognition of internal initiatives would move beyond the acknowledgement of a local community's role as defined by others towards the recognition of the community as it defines itself. The activity of the community is recognized and supported, rather than its potential contribution to some goal or project defined by an external actor. Organizational forms that differ from classical hierarchical models may also be supported if employed in successful examples of internal initiatives. The long-term work of non-state actors to support internal initiatives may also become more visible, along with needs for different funding models. Equally, all of this feeds into the view of benefit-sharing as a process: by recognizing internal initiatives, the CBD arguably makes space for views of benefit-sharing as ongoing negotiations, for example through community protocols (see Chapter 5: The view from below, for further discussion).

However, the numbers of texts assigned the category of 'internal initiative' are not high in comparison to those for the recognition and participation categories. They are nevertheless consistently higher than those for 'valorization of other worldview'. There are two peaks in the counts for texts supporting internal initiatives, in COP 7 and COP 12. In COP 7 they are for the most part related to decisions within the cross-cutting issues of mountain biological diversity and protected areas, and focus on community-based management. Before the second peak in COP 12, a dip is recorded in COPs 9 and 10. This is surprising for the case of COP 10, when the Nagoya Protocol, which recognized community protocols in international law, was adopted (Convention on Biological Diversity, 2010a). As noted, community protocols are feted in the literature as potential sources for pursuing local community issues at multiple levels – national, regional, and international (Jonas, Bavikatte and Shrumm, 2010; Tobin, 2013; Bavikatte, 2014; Bavikatte, Robinson and Oliva, 2015).

Considering the second peak in COP 12 places this in context. In COP 12, texts mentioning internal initiatives mostly fall within the cross-cutting issue of traditional knowledge. COP 12 marks an interesting turning point in the CBD's view of indigenous peoples and local communities. It was at this meeting that the term 'indigenous *peoples* and local communities' was adopted – prior to this meeting the CBD referred to 'indigenous and local communities'. The change in vocabulary followed years of tense discussions among the parties about the 2007 United Nations Declaration on the Rights of Indigenous Peoples (UNDRIP), and whether to adopt its language by referring to indigenous *peoples*. The fact that these debates took place in COPs 9 and 10, and that they concerned questions of national sovereignty given the connotations of self-determination attached to the use of the word 'peoples', goes some way to explaining the lack of text supporting internal initiatives. The decision to refer to indigenous *peoples* and local communities was taken at COP 12 (CBD COP Decision XII/12). Although a statement noting that the legal meaning of Article 8(j) would remain unchanged accompanied this shift in wording, the simultaneous peak in references to internal initiatives at the same meeting could indicate some discursive shift as one outcome of the years of debate. Specifically, the higher number of texts assigned the internal initiative category arguably chimes with the language used in the UNDRIP, which highlights, for example, the importance of 'control by indigenous peoples over developments affecting them and their lands, territories and resources', and 'that respect for indigenous knowledge, cultures and traditional practices contributes to sustainable and equitable development and proper management of the environment' (United Nations, 2007).

The discursive space opened up by the recognition of internal initiatives thus appears to be small, but may be growing. This would vindicate claims that internal initiatives such as community protocols would spread after their adoption in the Nagoya Protocol (Bavikatte, 2014). This small space appears more important than the category of valorization of other worldviews in terms of numbers alone. However, both of the categories suggest that contextual considerations can uncover the existence of meaningful shifts in the texts of COP decisions even where such language is rather rare. Overall, for these two discourse categories of 'other world valorization' and 'internal initiative', lower numbers of occurrences may well be linked to their higher political sensitivity, whereby they bring clear ramifications for national-level politics and policies. Recognizing the importance of traditional knowledge and customary sustainable use and talking about participation may equally lead to policy changes, but by encouraging internal initiatives and acknowledging the validity of other worldviews parties arguably make political statements about the rights of indigenous peoples and local communities with knock-on effects for policies including, for example, natural resource management, or land tenure (Parks, 2018).

Discourse categories that confirm the critical literature

The remaining categories used in the discourse analysis of the CBD's COP decisions are drawn from the more critical literature on global environmental governance (see the introduction to this chapter and Chapter 1: Local communities and global environmental governance). A strong presence of these discourse categories in the texts would indicate a lack of space for the common themes of importance to local communities. However, where some of the discourse categories derived from the critical literature appear alongside those drawn from the more optimistic literature, a discursive space for dialogue between different viewpoints may be opened up. The discourse categories of external imposition and exclusion can illustrate this idea. The former, exclusion, is a clear-cut case of a discourse category that closes spaces for local voices: indigenous peoples and local communities do not appear amongst stakeholders in the CBD in texts assigned this category. The category of external imposition shows how combinations can open up discursive spaces. In texts assigned the external imposition category, indigenous peoples and local communities are not involved in a decision, but are named as playing a role in its application in some way. The literature on CBNRM underlines the problems associated with this type of imposition. However, in some situations local communities are able to take externally imposed decisions and use them to their advantage. Whether such actions can then be used as a basis for a community to establish itself as a stakeholder and bring local issues to the attention of different authorities is beyond the scope of the present analysis, but the notion underlines the potential for subversive action within this discourse category. Where a text is assigned external imposition alongside participation, the scope for such action could be widened. Figure 4.5 shows the four discourse categories that draw from the critical literature for the sake of clarity.

Figure 4.5 shows that 'external imposition' is the most common among the four across all of the COPs, with 197 texts assigned this category. Because the sample of texts used for the discourse analysis is limited to those that are directly relevant to indigenous peoples and local communities, external imposition is an important indicator of limited spaces for expressing local viewpoints. Even where texts are assigned discourse categories that challenge the critical view of global environmental governance, the simultaneous presence of 'external imposition' means that the recognition, participation, or other positive aspect is curtailed: the local community is seen as a passive recipient of a decision or a policy made elsewhere despite any elements of recognition.[7] Investigating where the category of external imposition appears alongside other discourse categories is thus important to take into consideration.

Figure 4.6 shows that of the 197 texts assigned the category of 'external imposition', the majority, 134, are also assigned at least one category amongst the four that challenge the critical literature in some way. This means that some of the discursive spaces discussed previously are curbed.

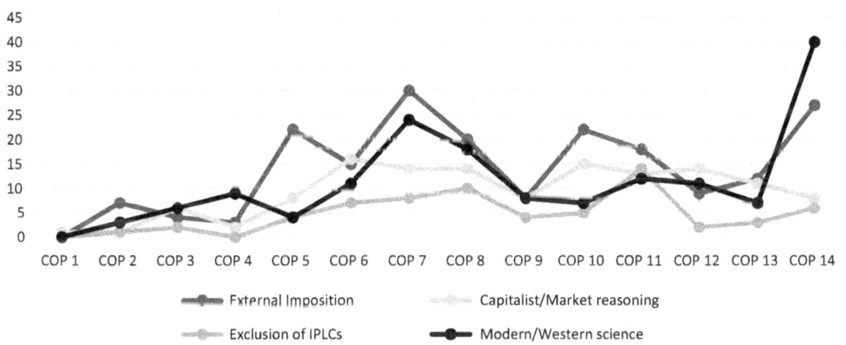

Figure 4.5 An overview of the discourse categories drawn from the critical literature.

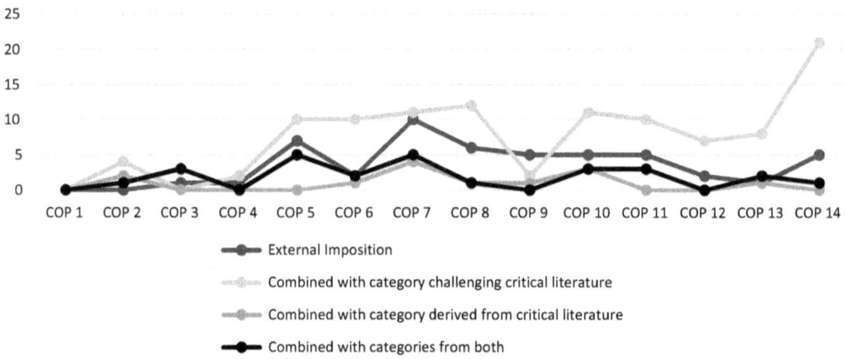

Figure 4.6 External imposition - combinations with other discourse categories over time.

However, the numbers of texts assigned discourse categories that are argued to create opportunities for local voices are much higher compared to those singled out in Figure 4.5. In addition, the external imposition categories combined with those that challenge the critical literature are fairly evenly spread. This shows that the curbing effect of external imposition on the discursive spaces for local voices in the CBD is limited. Even in COP 14, where 22 of these limiting combinations affect texts about local community participation, the overall number of texts on participation total 169. Although the external imposition discourse categories do limit discursive spaces, the general picture conveyed in Figure 4.1 holds.

The category of 'exclusion' is the least frequent in the discourse analysis as a whole. However, as mentioned in the introduction to the chapter, these relatively low numbers (particularly in comparison to the categories of participation and recognition) may be misleading, since the sample used for the analysis is limited to texts relevant to indigenous peoples and local communities. Many of the texts that exclude these groups may not thus be captured using this sampling technique. In other words, COP decisions that do not refer to indigenous peoples and local communities at all are likely to be those that exclude them, and limit the discursive spaces for raising themes of local importance. Further research covering all the entirety of the texts of COP decisions is necessary to explore this point in proper detail (Parks, Peters and Lennan, 2019). In addition, the 'exclusion' category was only assigned to texts that failed to include indigenous peoples and local communities in some kind of participation in the CBD that was explicitly extended to other actors. This strict application may mean that exclusion is underestimated in the analysis. To provide some more information and a preliminary idea about exclusion that moves beyond the method applied in this analysis, Figure 4.7 shows the frequency of texts that

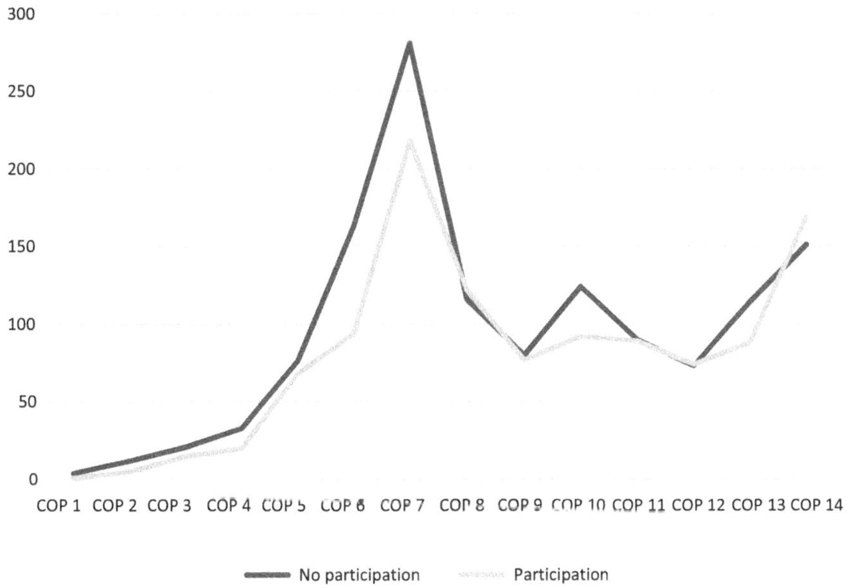

Figure 4.7 Texts assigned participation vs. texts not assigned participation over time.

are *not* assigned the counterpart category for exclusion: participation. The absence of participation does not, of course, equate to exclusion. Nevertheless, paying attention to the proportions of texts that refer to participation and those that do not provides information about its relative importance.

Figure 4.7 shows that the numbers of texts that do not refer to participation by indigenous peoples and local communities outnumber those that do until COP 8. From this point, the numbers begin to mirror one another more closely, with the exception of COP 10. In COP 14, the most recent meeting at the time of writing, the proportions are reversed. This suggests that while the CBD has tended to talk about participation for indigenous peoples and local communities, this was not a general feature of texts relevant to these groups until after COP 7. Nevertheless, the differences in the numbers of texts referring to participation and those that do not are relatively small. This suggests that at least where the CBD talks about indigenous peoples and local communities, the finding that exclusion is not a characteristic of this talk holds. The distribution of texts that do not mention participation by indigenous peoples and local communities is also more evenly spread amongst cross-cutting issues, compared to the concentration of texts that do mention participation within the cross-cutting issue of traditional knowledge.

The correspondence of texts assigned the exclusion category with the CBD's cross-cutting issues allows a little more information to be gleaned.

As noted, the majority of texts about participation fall within the cross-cutting theme of traditional knowledge. If texts that exclude indigenous peoples and local communities tend to appear in different cross-cutting issues, this would be a first indication that exclusion is underway elsewhere, and may impact discursive spaces where local themes can be raised to a greater extent than suggested by Figure 4.5.

Figure 4.8 shows that the exclusion category goes against the more general trend of confinement to the cross-cutting issue of traditional knowledge. The lengthy list of cross-cutting issues and the scattered distribution of texts among them, particularly after COP 6, suggests that exclusion may well be more prevalent in cross-cutting issues that are not represented to the same extent as traditional knowledge in the sample upon which this analysis is based. The need for further research on the matter is thus confirmed, and the caveat about the findings on small but growing discursive spaces for themes important to local communities is clear. These discursive spaces remain, but their limitations are underlined by the potential exclusion of local communities and indigenous peoples from the discourses that characterize COP decisions outside the cross-cutting issue of traditional knowledge.

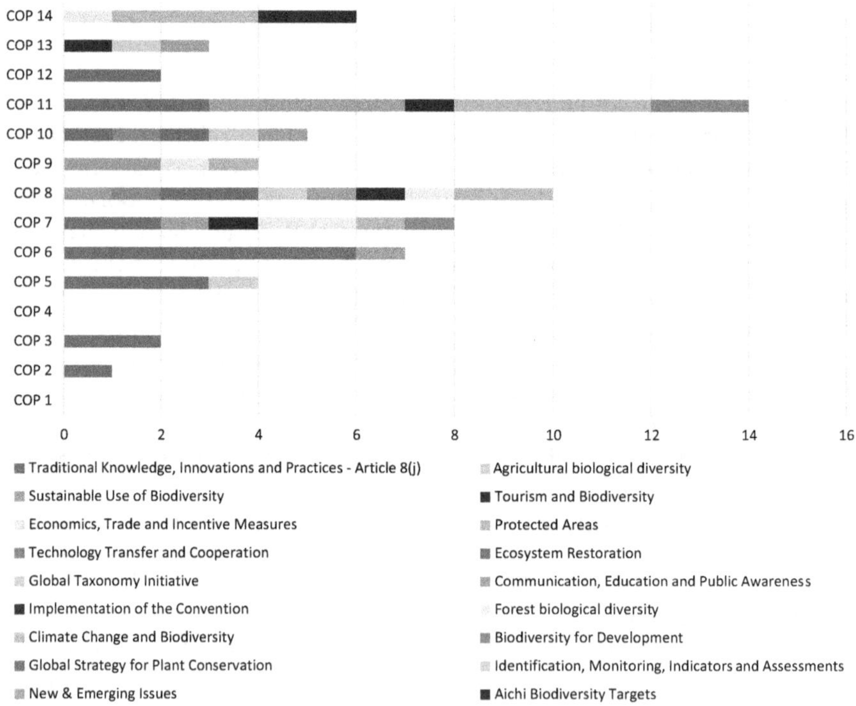

Figure 4.8 Exclusion of indigenous peoples and local communities - distribution across cross-cutting issues over time.

The remaining two discourse categories that draw on the critical literature are 'modern or western science' and 'capitalist or market reasoning'. Once more, the actual impact of these categories in terms of curtailing spaces for local issues depends on whether they are assigned alone or in combinations with categories that challenge the critical literature. In particular, where they appear alongside their counterpart categories – i.e. modern or western science is assigned to a text alongside the recognition of traditional knowledge, and capitalist or market reasoning is assigned alongside the valorization of other worldviews – it can be argued that the impact on spaces available for local themes is positive. The reasoning here is that texts that recognize these discourse categories alongside one another recognize different discourses as potentially complementary, rather than necessarily opposed. By recognizing different discourses, these texts can provide grounds for them to enter into dialogue. This is, perhaps, easier to argue for the categories modern or western science and traditional knowledge. These different sources of knowledge can complement and bolster one another, leading to broader understandings of and consequent solutions to environmental problems. The argument is more complex for capitalist or market reasoning and the valorization of other worldviews. Here, some argue that these worldviews are opposed in intractable ways (e.g. Natarajan and Khoday, 2014), while others see scope for accommodating the two in a single discursive framework of sustainable development (see Dryzek, 2005; UNEP, 2011). On this point, the analysis can provide an overview of the frequencies with which these two discourse categories appear alongside one another. This does not of course resolve the question in either direction, but does indicate whether attempts to integrate different types of knowledge and worldviews are made in CBD decisions.

Figure 4.5 shows that the numbers of texts assigned the discourse categories of 'modern or western science' and 'capitalist or market reasoning' are relatively low, particularly when compared with recognition and participation. Figure 4.9, on the other hand, shows that texts assigned the pairs of discourse categories that include these are even more rare. Of the two combinations, modern/western science and the recognition of traditional knowledge, and capitalist or market reasoning and the valorization of other worldviews, the former are more common. In addition, Figure 4.9 illustrates that a large proportion of texts assigned the modern/western science category are also assigned that of the recognition of traditional knowledge. The highest points for texts recording this combination are during COP 7 (for the most part on questions about the protection and retention of traditional knowledge and protected areas) and COP 8 (on island biodiversity and the status of traditional knowledge), as well as COP 14 (on climate change and biodiversity, marine and coastal biodiversity and the Rutzolijirisaxik Voluntary Guidelines for the Repatriation of Traditional Knowledge). The questions referred to in the texts assigned these combinations of discourse categories support the argument that they create spaces

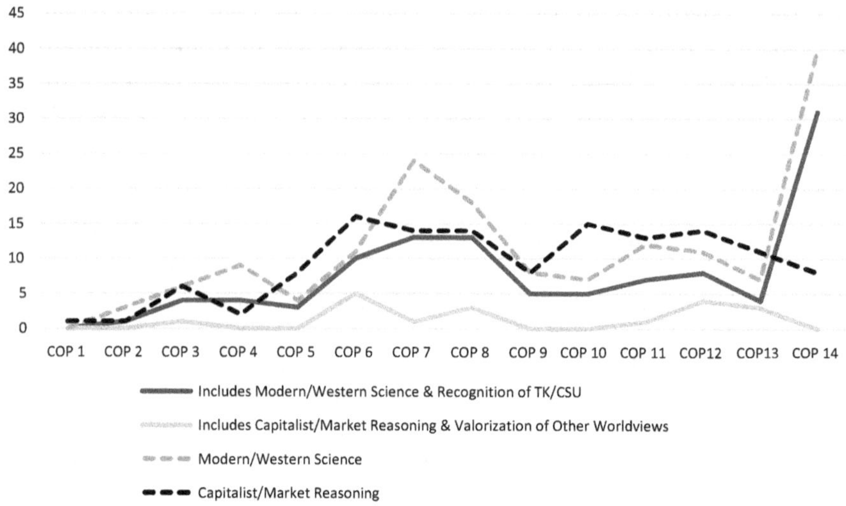

Figure 4.9 Discourse category combinations including Modern/Western Science & Recognition of Traditional Knowledge or Customary Sustainable Use, and Capitalist/Market Reasoning & Other Worldview.

for dialogue between different sources of knowledge. These combinations can thus be considered as adding to the small discursive spaces available for local themes. In particular, the idea of dialogue echoes many of the points discussed regarding recognition, both within and without the community, procedural guarantees, and the view of benefit-sharing as a long-term process. Communities pointed to the importance of mutual understanding and meaningful dialogue on these points. By providing discursive spaces that name different types of knowledge as valid to solving problems in the conservation of biodiversity, the texts assigned with this combination arguably provide a stronger form of recognition than that conferred in texts assigned the recognition of traditional knowledge category alone.

Combinations of the categories 'capitalist or market reasoning' and 'valorization of other worldviews' are even less frequent, representing only a small proportion of all the texts assigned the 'capitalist or market reasoning' category. This lends some evidence to the argument that it is particularly difficult to reconcile capitalist or market-based worldviews with those that are not based on dualistic views of nature and culture (e.g. Eastwood, 2019). Indeed, as the qualitative overview of texts assigned the 'valorization of other worldviews' category in the previous section reveals, it has taken many years for CBD COP decisions to begin to express the complexity of different worldviews, and this is still a rare occurrence. The generally low numbers of texts that refer to indigenous peoples and local communities and are explicitly based on capitalist or market reasoning does suggest some awareness that this type of approach may be less applicable to questions of interest to these groups. Overall, however, the question of worldviews – whether market-based or otherwise – is not broached to any clear extent in CBD COP decisions. Identifying the worldviews that underpin the decisions may be better located using other types of analysis, or indeed by using other methods such as interviews with negotiators, or observation at meetings. Literature on global environmental governance based on these approaches underlines the hidden and deeply ingrained nature of capitalist or market-based worldviews (e.g. Eastwood, 2019; Marion Suiseeya and Zanotti, 2019).

A final view of the discourse analysis is informative about where discursive spaces for local themes are located. Given that the bulk of the texts analyzed fall within the cross-cutting issue of traditional knowledge, how does the overall picture presented in Figure 4.1 change if texts that fall within other cross-cutting issues are removed? Do the recognition and participation categories continue to dominate, or do other categories begin to come to the fore? Figure 4.10 shows the distribution of the various discourse categories for texts within the cross-cutting theme traditional knowledge –1384 texts out of a total 3221.

The overall picture in Figure 4.10 shows that the discourse categories of recognition of traditional knowledge and customary sustainable use and participation for indigenous peoples and local communities continue to stand out. Nevertheless, the gap between these and other discourse

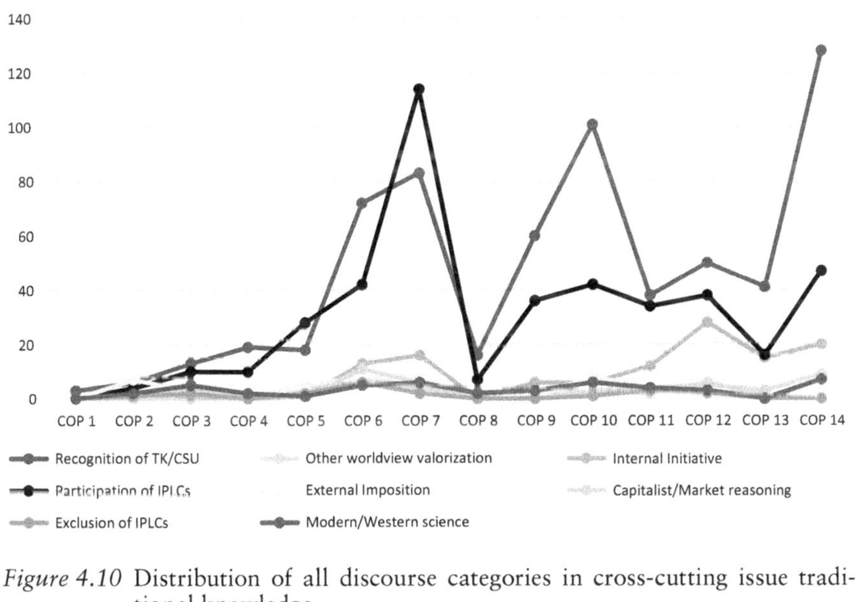

Figure 4.10 Distribution of all discourse categories in cross-cutting issue traditional knowledge.

categories is now less striking compared to the analysis as a whole. The internal initiative category also emerges a little more strongly within this cross-cutting issue, as others shrink. The most prominent change however is in the distribution and showing of the participation discourse category, which is much less frequent than the recognition category compared to the picture presented in Figure 4.1. The recognition of traditional knowledge and customary sustainable use is thus marked out as the strongest single category within the traditional knowledge issue. This is hardly a surprising result given that this is the 'natural' home for such statements to emerge, based on article 8(j).

Figure 4.10 also reveals an interesting trend that emerges less strongly in the overall view compared to this focus on the cross-cutting issue of traditional knowledge, namely the dramatic drop in all discourse categories in COP 8. This is particularly striking for the most frequent categories of recognition and participation, though internal initiative also drops from a previous upward trend. The explanation for this drop may lie in COP 7. This meeting, as mentioned, was a high point for talk about indigenous peoples and local communities, and also saw the adoption of the Akwé: Kon Guidelines on cultural, environmental and social impact assessments (Convention on Biological Diversity, 2004b) and the Addis Ababa Guidelines on sustainable use of biodiversity (Convention on Biological Diversity, 2004a). The drop in COP 8 following this high point suggests that parties to the CBD may have experienced some kind of 'wake-up call' about the extension of recognition to indigenous peoples and local communities, and drew back from this subject in the following COPs, which also coincided with the final debate and adoption of the UNDRIP. The UNDRIP certainly served to heighten sensitivities around the self-determination and rights of indigenous peoples during this period (Parks and Schröder, 2018). Though recognition returns to its previous levels at later COPs, with another clear peak in COP 10 due to the adoption of the Nagoya Protocol on Access and Benefit-sharing (Convention on Biological Diversity, 2010a) and the Tkarihwaié:ri Code of ethical conduct (Convention on Biological Diversity, 2010b), the same does not apply for participation, which only returns more strongly in COP 14.

Participation makes a stronger showing in the general analysis (see Figure 4.1) than within the cross-cutting issue of traditional knowledge (Figure 4.10). This indicates, as suggested earlier, that it is more evenly spread among different cross-cutting issues. Nevertheless, Figure 4.10 shows that participation drops rather dramatically in its 'natural home' of traditional knowledge following a high point in COP 7. Different interpretations are possible to explain this, including the argument made in the previous paragraph that parties to the CBD drew back from talk supportive of indigenous peoples and local communities in this period. Under the subsection Discourse categories that challenge the critical literature, however, it was argued that the participation discourse category appears to be generally accepted in the CBD, as suggested both by the high numbers of texts that refer to it and its spread into different cross-cutting issues. The argument about shying away from participation for indigenous peoples and local communities appears less convincing in light of this. The fact that mentions of participation are less frequent within the traditional knowledge cross-cutting issue could indicate instead this same general acceptance. If this is true, repeated calls for participation become superfluous within the traditional knowledge cross-cutting issue. Alternatively, the drop-off after COP 7 could suggest a less positive picture, namely that participation has become generally accepted but only because it is devoid of real meaning. Thus, other types of discourse categories, such as the recognition of traditional knowledge and customary sustainable use, remain prominent within the traditional knowledge cross-cutting issue because they are considered more effective for indigenous peoples and local communities. This interpretation echoes recent research on individual parties to the CBD that finds rhetorical support for the participation of indigenous peoples and local communities during COPs often fails to translate into support for their participation in national processes (De Chastonay, 2018). The question of what participation really means in the context of CBD COP decisions is discussed in detail in Chapter 5 (The view from above).

All very well, but what about everyone else?

The discourse analysis of CBD COP decision texts relevant to indigenous peoples and local communities suggests there is a small, delineated space, established in particular in COP 7, where the common themes that emerged from the case study comparison could be raised. Although drops in discourse categories of use to indigenous peoples and local communities are recorded after COP 7, the decisions from this meeting stick – COP decisions are cumulative, with previous decisions built upon by later ones. Two of the discourse categories that challenge the critical literature, the recognition of traditional knowledge and customary sustainable use and participation, make strong showings and can be linked to the common themes considered important by the local communities presented in this book. The discursive spaces for these themes are also limited in important ways. First, many of the texts fall within the cross-cutting issue of traditional knowledge, hinting at a kind of 'ghettoization' (Parks, 2018). There is some exception to this trend for the 'participation' discourse category, though a question mark remains as to the real meaning of participation. This is taken up in Chapter 5 (The view from above). Second, the discourse categories that are arguably more meaningful for common themes from the case studies, on internal initiatives and the valorization of worldviews not based in capitalist or market logic, are not very frequent in the CBD COP decisions. Third, exclusion poses a serious potential challenge that would curtail and rob power from this small discursive space. The analysis presented here has some limitations in this sense, and can only hint at what exclusion may be taking place in those areas of the COP decisions that do not refer directly to indigenous peoples and local communities. The previous section did highlight some evidence that exclusion is not underway so far as the texts in the sample considered are concerned. However, the limited data presented do suggest that exclusion – if present – does lie elsewhere in COP decision texts, at least in the discursive sense.[8]

This begs the question posed in the title of this section. It is all well and good that COP decisions leave some spaces for local themes, but what about the spaces accorded to other actors, and the discourses that characterize the COP decisions as a whole? Is there a specific discourse that has developed around indigenous peoples and local communities that builds a unique space for them to raise issues, or do other groups, such as industry actors, have similar and more powerful spaces to move in? To begin to address these points, a look at the proportion of space devoted to issues relevant to indigenous peoples and local communities in CBD COP decisions is useful.

Table 4.1 shows that the proportion of text devoted to indigenous peoples and local communities in each COP decision increases up to COP 7. The drop that follows COP 7, shown in Figure 4.1 and noted in the discussion of the distribution of texts within the traditional knowledge cross-cutting issue, is also put into context in Table 4.1. There is indeed a decrease of more than half in the numbers of texts relevant to indigenous peoples and local communities in COP 8, yet the proportional decrease is less striking at around 10 percent. Following COP 7, the proportion of texts relevant

Table 4.1 Proportions of COP decisions identified as referring to indigenous peoples and local communities

Conference of the Parties (COP)	*Texts (Paragraph or sub-paragraph) linked to IPLCs*	*Percentage (%) of text in COP decision referring to IPLCs*
1 (1994)	5	1.8
2 (1995)	17	6.7
3 (1996)	37	7.1
4 (1998)	53	9.7
5 (2000)	144	21
6 (2002)	257	18.7
7 (2004)	499	25.9
8 (2006)	238	15.2
9 (2008)	157	14.6
10 (2010)	216	17.5
11 (2012)	179	17.8
12 (2014)	147	1.2
13 (2016)	202	13.8
14 (2018)	320	20.2

to indigenous peoples and local communities remain at around 15 percent, with the exception of an increase in COP 14 with the adoption of the Rutzolijirisaxik Guidelines on the repatriation of traditional knowledge (Convention on Biological Diversity, 2018) and a rather dramatic decrease in COP 12. This drop is explained in part by the fact that about a quarter of the texts in COP 12 are devoted to the review of progress on the Strategic Plan for Biodiversity 2011–2020 and the Aichi Biodiversity Targets, and the COP decisions only focus on the involvement of actors following these reviews. Nevertheless, a sharp drop in the space devoted to issues relevant to indigenous peoples and local communities remains clear. Proportions increase again in COPs 13 and 14 however, suggesting this is a 'blip' rather than a trend. The proportion of space given over to matters of relevance for indigenous peoples and local communities appears quite substantial. This suggests that there is a specific discourse developing around these groups. The previous discussions in this chapter give an idea of some of the content of this discourse, summarized above as a small, but generally stable, space for raising local themes since COP 7.

Whether or not separate discourses apply to other actors within COP decisions is a question that is beyond the scope of this analysis, and further research is needed to look at other actors. The CBD addresses a range of other actors, including other multilateral bodies, international organizations, its parties, sub-national authorities, academia, non-governmental organizations, and industry.[9] Industry is perhaps the most interesting actor to contrast discursively with indigenous peoples and local communities, since the interests of these two broad groups of actors are often assumed to be opposed, as summarized previously in the discussion about competing worldviews. Though still at the very preliminary stages, some evidence about the language applied to industry actors is available. This evidence draws on research using an automated method of natural language processing which builds on the findings from the discourse analysis presented and its qualitative method for coding (see Parks, Peters and Lennan, 2019).[10] In this approach, the conceptual characterization associated with indigenous peoples and local communities in the COP decisions (including all of the texts of the decisions as opposed to the sample used in the analysis presented in this chapter) is contrasted with that connected to industry. The extent of overlap between the actor groups, as well as unique conceptual framings for each, are identified, thus providing some initial information about the potential existence of distinct discourses.

On a general level, the evidence from this analysis shows that the language connected to each of the groups – indigenous peoples and local communities on the one hand, and industry on the other – is strikingly similar. For example, 163 words or strings of words are connected with both groups on more than 100 occasions across the COP decisions. In contrast, just 6 words or strings are linked to industry alone, and the highest frequency across all COPs among these is 26 for the word 'appropriation'. Some 289 words or strings are instead linked to indigenous peoples and local communities alone, and 11 of these occur more than 50 times throughout the COPs. This suggests that there is indeed a distinct discourse about indigenous peoples and local communities, but also that there is no equivalent distinct discourse about industry actors. Indigenous peoples and local communities appear to be a privileged addressee of the CBD from this preliminary evidence. Further research is required to explore and refine these results and to consider the different words and strings linked to different descriptors for each actor in more detail; for example, the group 'indigenous peoples and local communities' includes the descriptors local communities, indigenous communities, indigenous peoples, representatives of indigenous peoples and others, and each carries different connotations. Further research can also consider the qualitative content of the words and strings to build on the analysis presented here. Nevertheless, the preliminary impression from this evidence bolsters the view of a small and stable discursive basis for indigenous peoples and local communities to raise issues of local importance, including the common themes that emerged from the case study comparison (see Chapter 3). In addition, it suggests that this small space is unique to the extent that no such space emerges in connection to industry actors.

This chapter took the common themes that emerged from the comparison of the case studies outlined in Chapters 2 and 3 and sought to examine whether there were discursive spaces available for them at the international level within the CBD. The aim was to assess whether the CBD is more closed off to local voices, as argued in some of the literature (e.g. Brand and Vadrot, 2013; Marion Suiseeya, 2014; Natarajan and Khoday, 2014; Eastwood, 2019), or whether it is more open, particularly in comparison to other arenas of global environmental governance (e.g. Reimerson, 2013; Bavikatte, 2014; Affolder, 2017; Schabus, 2017). The analysis demonstrated that a discursive space does exist, albeit curtailed in many ways. The following chapter builds on this general finding with a more pragmatic discussion about *how* local communities might be able to take advantage of this small space – and thus perhaps push its boundaries. The issue is approached from two different angles. First, the top-down angle: how does the CBD conceive of the participation of local actors? Further analysis of the discourse category 'participation' is carried out to respond to this question. The second angle is bottom up, where more space is left for the choices and agency of local actors. The discussion builds on the internal initiative discourse category to reflect on the potential of community protocols as bottom-up tools for conveying local themes to the international level in addition to their role in brokering relationships closer to home.

Notes

1 The discourse analysis is developed in Parks (2018). The figures in this chapter draw on the same data, albeit with the addition of data from COP 14, which took place in 2018. The data are available in the online data appendix at www.routledge.com/9780367181871. Mika Schröder provided invaluable assistance to complete the analysis and contributed reflections on the results.
2 See https://www.cbd.int/information/parties.shtml (accessed 18 July 2019).
3 In particular, the discourse analysis was carried out by two researchers following a codebook, and inter-coder reliability tests were carried out (Parks, 2018). In addition, the data for the analysis are available to interested readers at www.routledge.com/9780367181871.
4 See https://www.cbd.int/programmes/ (accessed 17 July 2019). Here, cross-cutting issues and thematic programmes on major biomes are referred to simply as 'cross-cutting issues'.
5 This discussion of the qualitative content of the 'other worldview valorization' category draws from Parks (2018).
6 This is reflected in the fact that in COP 6 half of the texts assigned the category 'other worldview valorization' are also assigned as 'capitalist/market reasoning'.
7 Though not necessarily canceled out, as already argued previously.
8 Exclusion here is considered as a discursive category where indigenous peoples and local communities are not considered in COP decision texts while other types of actors are. Exclusion can also be understood in a much wider sense, including more hidden forms, for example at the stage of the implementation of decisions, but these are beyond the present scope.
9 See https://www.cbd.int/cooperation/about/partners.shtml (accessed 20 July 2019).
10 Wim Peters produced the data commented on in brief here, and Mitchell Lennan assisted in preparing the analysis and interpreting the preliminary results.

References

Adger, W. N. *et al.* (2001) 'Advancing a political ecology of global environmental discourses', *Development and Change,* 32(4), pp. 681–715.

Affolder, N. (2017) 'Non-State Actors', in Morgera, E. and Razzaque, J. (eds) *Encyclopedia of Environmental Law: Biodiversity and Nature Protection Law.* Cheltenham: Edward Elgar, pp. 387–398.

Bavikatte, K. (2014) *Stewarding The Earth. Rethinking Property and the Emergence of Biocultural Rights.* Oxford: Oxford University Press.

Bavikatte, K., Robinson, D. F. and Oliva, M. J. (2015) 'Biocultural community protocols: dialogues on the space within', *IK: Other Ways of Knowing,* 1(2), pp. 1–31.

Birnie, P., Boyle, A. and Redgewell, C. (2009) *International Law and the Environment.* Oxford: Oxford University Press.

Brand, U. and Vadrot, A. B. M. (2013) 'Epistemic selectivities and the valorisation of nature: the cases of the Nagoya Protocol and the intergovernmental science-policy platform of the biodiversity and ecosystem services (IPBES)', *Law, Environment and Development Journal,* 9(2), pp. 202–220.

De Chastonay, A. C. (2018) *Like-Minded Megadiverse Countries, Community-Based Rights, and the International Governance of Biodiversity.* Université de Lausanne.

Convention on Biological Diversity (2002) *Report on the Status and Trends Regarding the knowledge, innovations and practices of indigenous and local communities relevant to the conservation and Sustainable Use of Biodiversity.* CBD Decision VII/12, Annex, Montreal.

Convention on Biological Diversity (2004a) *Addis Ababa Principles and Guidelines on the Sustainable Use of Biodiversity.* CBD Decision VII/12, Annex, Montreal.

Convention on Biological Diversity (2004b) *Akwé: Kon Voluntary Guidelines for the Conduct of Cultural, Environmental and Social Impact Assessment Regarding Developments Proposed to Take Place on, or which are Likely to Impact on, Sacred Sites and on Lands and Waters traditionally Occupied or Used by Indigenous and Local Communities.* CBD Decision VII/16, Annex, Montreal.

Convention on Biological Diversity (2010a) *The Nagoya Protocol on Access to Genetic Resources and the Fair and Equitable Sharing of Benefits Arising from their Utilization to the Convention on Biological Diversity.* CBD Decision X/1, Annex, Montreal.

Convention on Biological Diversity (2010b) *Tkarihwaié:ri Code of Ethical Conduct on Respect for the Cultural and Intellectual Heritage of Indigenous and Local Communities.* CBD Decision X/42, Annex. Montreal.

Convention on Biological Diversity (2016) *Mo'otz Kuxtal Voluntary guidelines for the development of mechanisms, legislation or other appropriate initiatives to ensure the "prior and informed consent", "free, prior and informed consent" or "approval and involvement", depending on national circumstances, of indigenous peoples and local communities for accessing their knowledge, innovations and practices, for fair and equitable sharing of benefits arising from the use of their knowledge, innovations and practices relevant for the conservation and sustainable use of biological diversity, and for reporting and preventing unlawful appropriation of traditional knowledge.* CBD Decision XIII/18, Annex, Montreal.

Convention on Biological Diversity (2018) *The Rutzolijirisaxik Voluntary Guidelines for the Repatriation of Traditional Knowledge of Indigenous Peoples and Local Communities Relevant for the Conservation and Sustainable Use of Biological Diversity.* CBD Decision XIV/6, Annex, Montreal.

Dryzek, J. S. (2005) *The Politics of the Earth: Environmental Discourses.* Second edition. Oxford: Oxford University Press.

Eastwood, L. E. (2019) *Negotiating the Environment: Civil Society, Globalisation and the UN.* London: Routledge.

Gellers, J. C. (2015) 'Greening critical discourse analysis: applications to the study of environmental law', *Critical Discourse Studies*, 12(4), pp. 482–493.

Hajer, M. and Versteeg, W. (2005) 'A decade of discourse analysis of environmental politics: achievements, challenges, perspectives', *Journal of Environmental Policy & Planning*, 7(3), pp. 175–184.

Harrop, S. R. and Pritchard, D. J. (2011) 'A hard instrument goes soft: the implications of the Convention on Biological Diversity's current trajectory', *Global Environmental Change*, 21(2), pp. 474–480.

Jonas, H., Bavikatte, K. and Shrumm, H. (2010) 'Community protocols and access and benefit sharing', *Asian Biotechnology and Development Review*, 12(3), pp. 49–77.

Jonas, H. C. (2017) 'The Evolution of Indigenous Peoples' and Community Conserved Territories and Areas (ICCAs) in International Biodiversity Law and Conservation Policy', in Morgera, E. and Razzaque, J. (eds) *Encyclopedia of Environmental Law: Biodiversity and Nature Protection Law*. Cheltenham: Edward Elgar, pp. 145–160.

Marion Suiseeya, K. R. (2014) 'Negotiating the Nagoya Protocol: indigenous demands for justice', *Global Environmental Politics,* 14(3), pp. 102–124.

Marion Suiseeya, K. R. and Zanotti, L. (2019) 'Making influence visible: innovating ethnography at the Paris Climate Summit', *Global Environmental Politics,* 19(2), pp. 38–60.

Morgera, E. and Savaresi, A. (2013) 'A conceptual and legal perspective on the green economy', *Review of European Community & International Environmental Law*, 22(1), pp. 14–28.

Morgera, E. and Tsioumani, E. (2011) 'Yesterday, today and tomorrow: looking afresh at the Convention on Biological Diversity', *Yearbook of International Environmental Law*, 21(1), pp. 3–40.

Moses, J. W. and Knutsen, T. L. (2012) *Ways of Knowing. Competing Methodologies in Social and Political Research*. Second edition. Basingstoke: Palgrave Macmillan.

Murphree, M. (2004) 'Communal approaches to natural resource management in Africa: from whence to where?', *Journal of International Wildlife Law and Policy*, 7, pp. 203–216.

Natarajan, U. and Khoday, K. (2014) 'Locating nature: making and unmaking international law', *Leiden Journal of International Law*, 27(3), pp. 573–593. doi: 10.1017/S0922156514000211.

Nelson, F. (2010) 'Democratizing Natural Resource Governance: Searching for Institutional Change', in Nelson, F. (ed.) *Community Rights, Conservation and Contested Land: The Politics of Natural Resource Governance In Africa.* London: Earthscan, pp. 310–333.

Parks, L. (2018) 'Spaces for local voices? A discourse analysis of the decisions of the Convention on Biological Diversity', *Journal of Human Rights and the Environment*, 9(2), pp. 141–170.

Parks, L., Peters, W. and Lennan, M. (2019) *Guidelines and Codes on the Participation of Indigenous Peoples and Local Communities of the Convention on Biological Diversity: A Comparative Analysis Using Natural Language Processing,* BENELEX Working Papers. 23. Available at: https://ssrn.com/abstract=3384691.

Parks, L. and Schröder, M. (2018) 'What we talk about when we talk about "local" participation in international biodiversity law. The changing scope of indigenous peoples and local communities' participation under the Convention on Biological Diversity', *Participation and Conflict,* 11(3), pp. 743–785.

Reimerson, E. (2013) 'Between nature and culture: exploring space for indigenous agency in the Convention on Biological Diversity', *Environmental Politics,* 22(6), pp. 992–1009.

Robinson, D. F. (2015) *Biodiversity, Access and Benefit-Sharing. Global Case Studies*. London: Routledge.

Schabus, N. (2017) 'Traditional Knowledge', in Morgera, E. and Razzaque, J. (eds) *Encyclopedia of Environmental Law: Biodiversity and Nature Protection Law.* Cheltenham: Edward Elgar, pp. 264–279.

Svarstad, H. (2004) 'A Global Political Ecology of Bioprospecting', in Paulson, S. and Gezon, L. L. (eds) *Political Ecology Across Spaces, Scales and Social Groups*. New Brunswick, NJ: Rutgers University Press, pp. 239–256.

Tobin, B. (2013) 'Bridging the Nagoya compliance gap: the fundamental role of customary law in protection of indigenous peoples' resource and knowledge rights', *Law, Environment and Development Journal*, 9(13), pp. 142–162.

Traditional Health Practitioners of Bushbuckridge (2010) *Biocultural Protocol of The Traditional Health Practitioners of Bushbuckridge.* Available at: http://community-protocols.org/wp-content/uploads/documents/South_Africa-Bushbuckridge_Biocultural_Protocol.pdf.

Uggla, Y. (2010) 'What is this thing called "natural"? The nature–culture divide in climate change and biodiversity policy', *Journal of Political Ecology,* 17(1), pp. 79–91.

UNEP (2002) *Bonn Guidelines on access to genetic resources and fair and equitable sharing of the benefits arising out of their utilization.* CBD Decision VI/24, Annex, Montreal.

UNEP (2011) *Towards a Green Economy: Pathways to Sustainable Development and Poverty Eradication – A Synthesis for Policy Makers.* Available at: www.unep.org/greeneconomy.

United Nations (2007) *Declaration on the Rights of Indigenous Peoples.* UN Resolution 61/295.

Wilkinson, K. (2014) 'Payment for "ecosystem services" and the "green economy": green-washing or something new?', *Journal of Human Rights and the Environment,* 5(2), pp. 168–191.

5 Local participation in international processes

Views from above and below

This chapter discusses some of the paths available for local themes to be brought to the attention of international actors, including the challenges they present. Chapter 4 discussed whether discursive grounds exist in the arena of the CBD for the themes that emerged from a comparison of the five local community case studies presented in Chapters 2 and 3. Although discursive grounds for local issues were found to be limited, one key finding in Chapter 4 was that CBD COP decisions often refer to the participation of indigenous peoples and local communities, in a variety of cross-cutting issues. If local communities can participate in international environmental governance, the issues they consider important could be taken up at the global level. Different communities may seek to participate in different ways: through 'insider' strategies within the existing environmental governance system, 'outsider' strategies that challenge or even reject the existing system, or a combination of both (Dryzek, 2005; Sikkink, 2005). Those communities that contribute to shape the international level of governance through their participation could also be argued to create opportunities for others on the basis of the (admittedly limited) evidence discussed in this book so far. If there are some commonalities among the themes that local communities consider important in their discussions around benefit-sharing, and if those themes are more generally important to local communities as far as questions of environmental governance are concerned, then the impact of local community participation at the international level, where decisions that affect communities across the globe are taken, could have a wider positive effect.

Though the impacts of participation are notoriously difficult to trace to single actors (Bosi, Giugni and Uba, 2016), and the evidence for commonalities among local experiences is limited, investigating paths of participation in global environmental governance for local communities is also important in a view of substantive democracy at the international level. There are many reasons why local community participation as one plank in a more substantive form of democracy at the international level is of interest. First, a number of multilateral environmental agreements explicitly recognize the important role that local communities play in conservation, including the

CBD and the UNFCCC (Reimerson, 2013; Savaresi, 2018). Soft law environmental instruments such as the Rio Declaration make similar observations (United Nations, 1992). Equally, literature on the local applications of environmental policies note that where local communities play an active part in developing and implementing environmental policy, effectiveness improves (e.g. Bäckstrand and Lövbrand, 2006; Child and Barnes, 2010). Second, intergovernmental action on environmental issues has proceeded in fits and starts, with various moments of blockage and a very questionable track record in terms of progress on key goals (including emissions reductions, limiting the increase of global temperatures, and conserving biodiversity). Environmental questions are also often inherently transnational, and require cross-border cooperation at all levels, including the local (Mitrotta, 2019). Many scholars have thus made calls for an ecological democracy that involves more substantive input from all citizens, including at the global level of decision-making (e.g. Bowman, 2010; Dryzek and Stevenson, 2011, see discussion in Chapter 6). By beginning a discussion about a more practical view of participation, this chapter seeks to open up reflections about the existing paths for local communities to take part in international decision-making in the CBD.

Two views of participation are explored: one from above, and one from below. Both focus on 'insider' strategies that can be used to open up spaces within the existing system. The view from above explores the range of descriptions of participation for indigenous peoples and local communities that exist in the CBD's COP decisions. These descriptions are drawn from the texts assigned the discourse category 'participation' in the analysis presented in Chapter 4. The participation category emerges as particularly important in CBD COP decisions, both in terms of the number of times it is raised and in terms of its spread across a range of cross-cutting issues over time. However, the category reveals little more than the fact that a text refers to some kind of participation of indigenous peoples and local communities – what form this participation will take, where it will take place, whether it is mandatory or simply hoped for – these and other questions remain unanswered by the analysis presented in Chapter 4. To investigate paths for local participation as they are described in CBD COP decisions afresh, an inductive analysis of texts assigned the participation category is presented here.

The view from below also builds on the discourse analysis presented in Chapter 4, but in a more reflective way. The focus is on the idea of 'internal initiatives', another discourse category presented in Chapter 4. Texts in COP decisions that encourage and support the initiatives of local communities to manage and protect natural resources were assigned this category. These initiatives are a version of bottom-up management, and the internal initiative category was assigned to texts that recognized and supported locally developed models. To explore how bottom-up internal initiatives could contribute to participation, the chapter discusses the potential of community protocols. These are tools used by local communities to outline their practices, institutions, customary law, rights, visions of development, and other information for protecting and managing natural resources. They are first and foremost tools that are used for building relationships between local communities and different external actors in domestic contexts. Nevertheless, their potential for feeding into international-level decision-making in the CBD has been raised (Tobin, 2013; Bavikatte, 2014), and is worth exploring in light of the findings that emerge from the top-down analysis in the following section. The discussion reflects on how community protocols – as just one form of community-led initiative among others – could be used as a bottom-up path for local participation in international processes in the CBD. In particular, it asks whether community protocols help address the dominant discourses identified in critical literature (see the overview in Chapter 1: Local communities and global environmental governance), the clear power imbalances between local communities and international arenas, and the gatekeepers between these levels.

The view from above – participation and local communities in the CBD

As noted, talk about the participation of 'indigenous peoples and local communities' is both frequent and relatively widespread in CBD COP decisions. Yet the discourse category used in Chapter 4 communicates no more specific information beyond this, and participation can take on many different meanings. This section presents further qualitative inductive analysis of all the texts assigned this discourse category to uncover the actual meanings attached to the idea of participation in CBD COP decisions where local communities are concerned.[1]

The texts are analyzed inductively as opposed to the deductive approach of the analysis in Chapter 4, where predetermined discourse categories were assigned to texts (see Chapter 4: Analyzing the discourses of the Convention on Biological Diversity). In this chapter, the same COP decision texts are used to generate more general descriptions of participation, which were developed until all the texts were covered. More than one description can be assigned to one text, however no clear patterns of combinations emerged.[2] This suggests that the descriptions developed are sufficiently differentiated. The analysis also draws on the literature on framing, where 'a particular definition is in charge of a situation' (Gamson, 1985: 686), and on Benford and Snow's (2000) distinction between prognostic, diagnostic, and motivational frames. Prognostic frames inform us about understandings of a problem. Diagnostic frames provide information about how a problem should be solved, and motivational frames urge action. Classifying descriptions of participation according to these three categories helps guide the discussion in a more structured way, and feeds into the overall goal of investigating framings of benefit-sharing (see the discussion in Chapter 1: Norm diffusion – an interdisciplinary overview).

The analysis also looks at which frames of participation correspond with which cross-cutting issues, and at the territorial levels at which participation is supposed to take place. Frames that foresee participation at the international level are of particular interest for the present discussion, as these provide some kind of direct path for local community participation. Identifying the relevant territorial levels at which different texts placed participation proved challenging, not least due to the variety of authorities referred to within States, and the number of texts with no explicit references. The system applied in the analysis is as follows:

1) Local level: this is used where participation takes place in local bodies and groups that are separable from the State (excluding municipal councils, for example) and is only assigned where there is an explicit reference. Participation through internal initiatives such as community protocols is also coded as local.
2) Domestic level: this is used where participation takes place with the involvement of a range of national and sub-national actors that are not separable from the State (including municipal councils, for example).
3) International level: this is used where participation takes place with the involvement of international actors such as international organizations.
4) Not specified: this is used where no explicit or clear reference to where participation is to take place is made.

In addition, the levels of participation can be combined to account for both the processes and the 'destinations' of participation. For example, a text asking parties to the CBD to gather information from local communities and to subsequently share that information with the CBD Secretariat or some other CBD body is assigned both 'domestic' and 'international' levels, but not a 'local' level since State bodies decide which information is included, from whom, and how it is presented. If the text asks for local communities to provide information directly to a CBD body, it is assigned as 'international'.

A total of 30 different descriptions of participation emerged from the 1144 texts, and were assigned 1529 times. Table 5.1 lists them, along with brief explanations, and groups them in line with framing concepts. The descriptions that emerge from the analysis do not correspond exactly to the Benford and Snow (2000) schema however. They are thus grouped according to whether they are: 1) diagnostic explanations of the need for local community participation; 2) prognostic descriptions of mechanisms for participation – both more and less detailed; 3) motivational/prognostic frames about who should be responsible for or have power over the arrangements for participation; or, finally, 4) prognostic/diagnostic frames about the limits of participation. A description referring to the presence of

Table 5.1 Descriptions of participation in CBD COP Decisions 1–14

1)	*GENERAL STATEMENTS*: General statements about the need to strengthen the involvement or participation of indigenous peoples and local communities (IPLCs) and/or the need to involve IPLCs.	*EMPOWER*: A statement that uses the specific word 'empower' in relation to IPLC participation.
	EXPERT: A statement about participation that underlines IPLCs as experts, or traditional knowledge as an important source of information.	*PARTICIPATION FOR RESPECT*: Statement about effective participation that is, or contributes to decisions that are, respectful, culturally appropriate, sensitive, non-intrusive, etc. Includes the idea that participation will be culturally dependent.
	WOMEN: Statement underlining the need for and/or importance of participation by IPLC women.	*YOUTH*: Statement underlining the need for and/or importance of participation by IPLC youth.
	UNDRIP: The text refers to the United Nations Declaration on the Rights of Indigenous Peoples.	
2)	*IMPLEMENTATION*: IPLCs are named or emphasized as actors in implementing the CBD in some way.	*INPUTS FROM IPLCs*: Invitations or statements encouraging IPLCs, amongst other actors, to provide information to a CBD body (e.g. a working group, the Executive Secretariat, etc.).
	GENERAL CALL FOR NOMINATION: A statement that names or encourages another actor to nominate IPLCs to some body or process of the CBD (not delegations).	*PARTIES SHOULD NOMINATE IPLCs*: A statement that encourages parties to the CBD to include IPLCs in their delegations.
	FUNDS: The text contains a statement about the need for funds to allow IPLC participation.	*INDIGENOUS ISSUES BODY*: The statement refers to participation via contacts with the United Nations Working Group on Indigenous Populations or another indigenous issues body or bodies where IPLCs participate.
	DEFINITE INVOLVEMENT ON SPECIFIC MATTER: IPLCs will definitely be involved in decisions on or a specific matter described. Includes statements about the involvement of the Ad Hoc Working Group on Article 8(j).	*ENCOURAGE INVOLVEMENT ON SPECIFIC MATTER*: IPLCs should be invited and/or encouraged to be involved in decisions on or a specific matter described.
	IPLC EXCHANGE: Support or call for exchanges between IPLCs (information, consultation, etc.)	*INCLUDE VIEWS*: Statement that views or information received from IPLCs will be reproduced in a report, decision, or other document.
	MEASURES TO IMPROVE PARTICIPATION: Statement about implementing specific measures to strengthen or improve IPLC participation.	*CAPACITY-BUILDING*: A statement about capacity-building, education, or awareness building for IPLCs, framed as a mode of participation.

(*Continued*)

CBD AS BASIS FOR PARTICIPATION CLAIM: The text invites or encourages IPLCs to use the CBD as the basis of or to strengthen requests for participation.

NETWORKS AS PARTICIPATION: IPLCs are named as actors in networks or partnerships through which they participate.

PARTICIPATORY MODELS: Statement calling for participation to take place through participatory models, implying meaningful dialogue.

3) *NAMED ACTOR SHOULD ENSURE PARTICIPATION*: A particular actor is named and should act to ensure that IPLCs can participate in decision-making, some other forum, or be able to take action in a specific sector.

CHOOSE REPRESENTATIVES: IPLCs themselves should choose their representatives for some body.

CONTROL: IPLCs have or should have control over some process.

NAMED ACTOR SHOULD SUPPORT PARTICIPATION: A particular actor should act in a way that supports IPLCs in their participation in decision-making, some other forum, or to take action in a specific sector.

INFORMATION ON PARTICIPATION: A request for information on IPLC participation.

4) *IMPROVE PARTICIPATION*: Text recognizing that more work is needed to improve IPLC participation in the CBD.

QUALIFIER: The text includes a qualifier on IPLC participation (e.g. 'as far as appropriate and possible' or 'in accordance with national legislation'.

LIMITS OF PARTICIPATION: Statement noting that levels of IPLC participation also depend on factors outside the control of the CBD.

---

a 'qualifier' on participation is included in this last group. This is always assigned in combination with others, and denotes some limit placed on the type of participation mentioned in the text.

To understand the relative significance of this rather long list of descriptions, Figure 5.1 shows the overall frequencies with which each appears in the texts of COP decisions.

Half of the descriptions are prognostic frames detailing mechanisms for participation, and together these account for more than half of the texts (55 percent). This focus on mechanisms reinforces the idea raised in Chapter 4 about the participation of indigenous peoples and local communities being generally accepted in CBD COP decisions. Some 20 percent of the remaining texts are diagnostic and underline the need for participation. This could be read as a challenge to the claim that there is general acceptance of the need for participation by indigenous peoples and local communities. However, more than a third of the texts here are general statements, which are often found in paragraph headings and simply recall previous decisions that established the need for participation by these groups. Another 21 percent of the texts frame power over participation processes. This is an

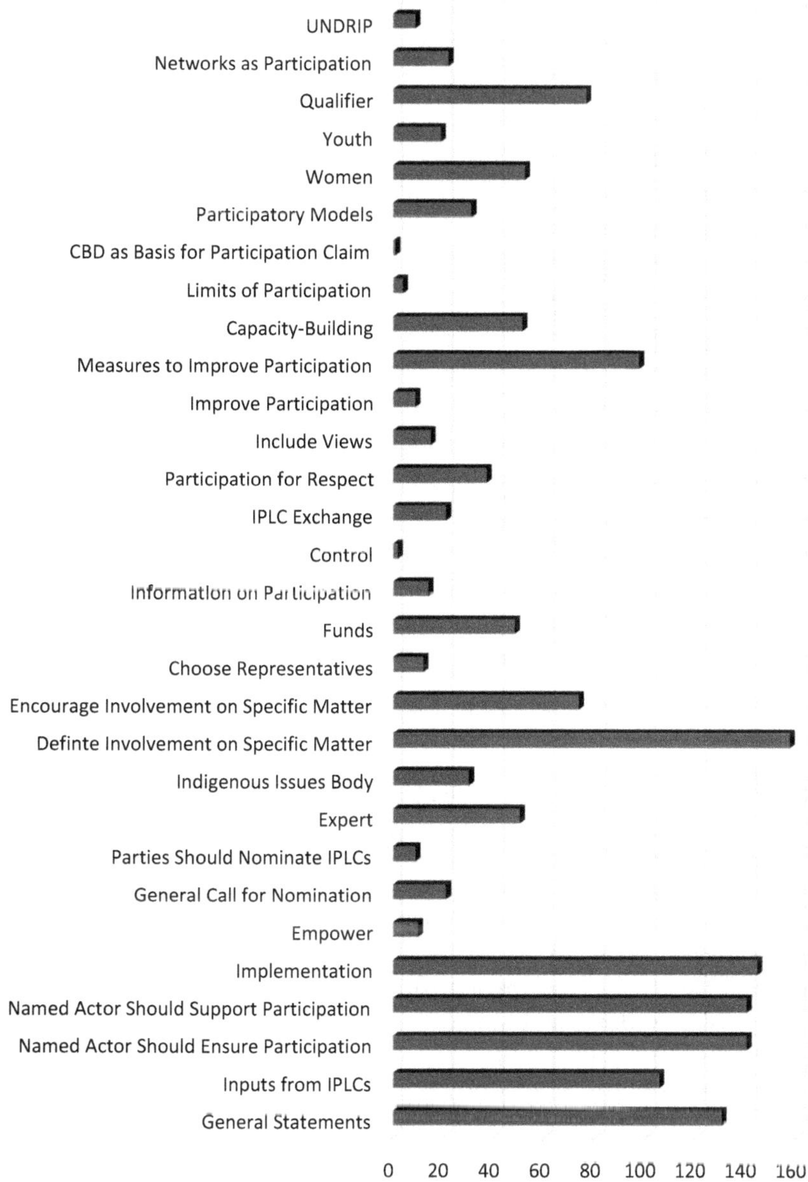

Figure 5.1 Distribution of participation descriptions.

important category, since assigning responsibility for participation makes a specific actor or group of actors ultimately accountable for failures (Parks and Schröder, 2018). Just 4 percent refer to the limits to participation, and amongst these are reflections about the inherent limits of participation, and the need to improve mechanisms.

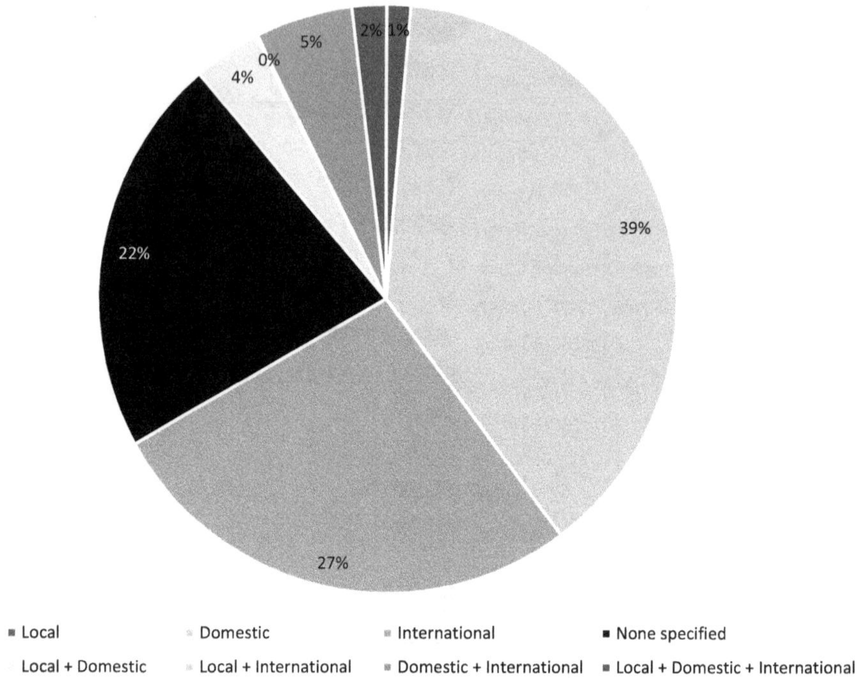

Figure 5.2 Levels of participation (all descriptions).

Regarding the levels where participation is to take place, Figure 5.2 shows that the domestic level (which includes local authorities as discussed in the introduction to this chapter) is the most common, and accounts for about 39 percent of all the texts. This rises to circa 50 percent in combinations that include the domestic level. Some 27 percent of texts refer to the international level, rising to 34 percent with combinations. About 22 percent of texts do not specify where participation should take place, and just 7 percent or so refer to the local level, again counting all relevant combinations. The domestic level remains the most important location for participation by local communities. Though interesting, this is not surprising given the continued centrality of States in the international system, and the fact that the CBD is a framework treaty that assigns the implementation of its decisions to its parties. The proportion of texts that refer to the international level as the sole location of participation by indigenous peoples and local communities appears quite sizeable in this light.

Table 5.1 points to a smaller group of 6 descriptions that occur more than 80 times each in the COP decision texts. Their frequency over time is shown in Figure 5.3. The add-on category 'qualifier' appears only 76 times, and is not therefore included in Figure 5.3. The distribution of qualifiers

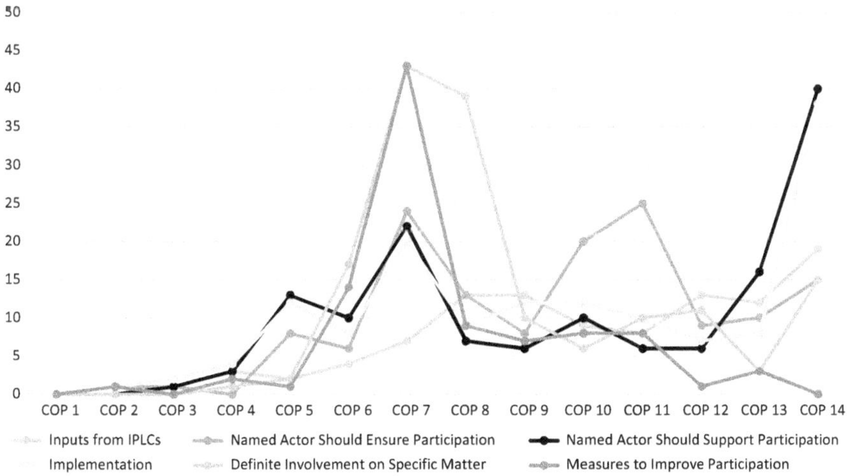

Figure 5.3 Descriptions of participation over time (descriptions used 80+ times).

to participation amongst the different descriptions does not reveal any distinct pattern, and qualifiers apply to many descriptions in low numbers. Thus, qualifiers do not indicate any significant brake on participation by indigenous peoples and local communities, and are not discussed further. After discarding 'general statements', which also appears more than 80 times in the analysis, for the reasons already mentioned, these 6 still account for just over half of the descriptions assigned (783). Amongst these, 4 are mechanisms for participation – 'inputs', 'implementation', 'definite involvement on a specific matter', and 'measures to improve participation' – while the remaining 2, 'named actor should ensure participation' and 'named actor should support participation', – concern power over participation processes.

Two of the most frequent descriptions, 'definite involvement on a specific matter' and 'measures to improve participation' are fairly similar in terms of distribution and content. Both suggest that participation will or has taken place, specifying the issue to be participated in on one hand, and highlighting the need to strengthen participation on the other. However, neither specifies any precise mechanism. A partial exception here is the inclusion of participation via the CBD's Ad Hoc Working Group on Article 8(j), which includes representatives of indigenous peoples and local communities and has been noted as a clear source for local community influence on CBD processes (e.g. Morgera, 2017) with the 'definite involvement' description. However, texts referring to the Ad Hoc Working Group, including references to decisions it has taken in the past, account for only 30 out of the 157 texts assigned this description and no clear pattern in combinations of descriptions emerges. Some of the less frequent descriptions of participation also peak in COP 7. Though assigned to different texts, they may provide clues about what 'definite involvement' and 'measures for improvement' might include. Among them, two are diagnostic and provide information about the motivation behind calls for participation. They are descriptions of local communities as 'experts' and texts that explain 'participation for respect'. The latter is particularly interesting in light of the discussions of common themes from the case studies, where respect was an important element in the central common theme of recognition as well as views of benefit-sharing as a process (see Chapter 3). The description only appears 37 times, and 10 of these are in COP 7. The remaining description that spikes at COP 7, 'capacity-building', provides a hint on the possible content of 'measures for improvement'. It underlines that local communities should benefit from capacity-building to bolster their participation in CBD processes.

Around half of the texts for the 'definite involvement' description – 82 – are concentrated in COPs 7 and 8, and a similar proportion of the texts for 'measures to improve participation' fall in COP 7 alone. This reinforces the view of a 'wake-up call' about local community involvement after COP 7 in the context of the adoption of instruments with strong elements of participation for indigenous peoples and local communities, as well as ongoing debates on the UNDRIP (see Chapter 4). This delimitation also affects the importance of these descriptions, limiting them to a certain time. Regarding the levels where participation is to take place, 'definite involvement' refers mostly to the domestic and unspecified categories for COPs 7 and 8. Similarly, the vast majority of texts assigned the 'measures to improve' description (31 out of 43) refer to the domestic level in COP 7. This also contributes to explain the relatively short-lived nature of these descriptions. Local community participation was a core theme in instruments such as the Akwé: Kon Guidelines (Convention on Biological Diversity, 2004), which refer to the participation of indigenous peoples and local communities in domestic processes including EIAs. This may have been interpreted as a precedent for interventions in domestic politics that later led States to back away from such language.

To investigate this further, Figure 5.4 compares the 'definite involvement' description with its conditional counterpart, 'encourage involvement on a specific matter'. The latter frames participation in an optional view, asking parties to encourage rather than ensure participation. Comparing the two gives an idea of where stronger and weaker commitments to participation appear. It should be borne in mind that less directive language could also be interpreted as underlining the need to provide support for participation, given the costs and investment required.

The comparison shows that the 'definite' description is more frequent overall (though concentrated in COPs 7 and 8). However, the 'encourage' version of the description gains ground over time, and is more common in COPs 13 and 14. The significance of this change in language is debatable, but the decrease in the use of more directive language does support the

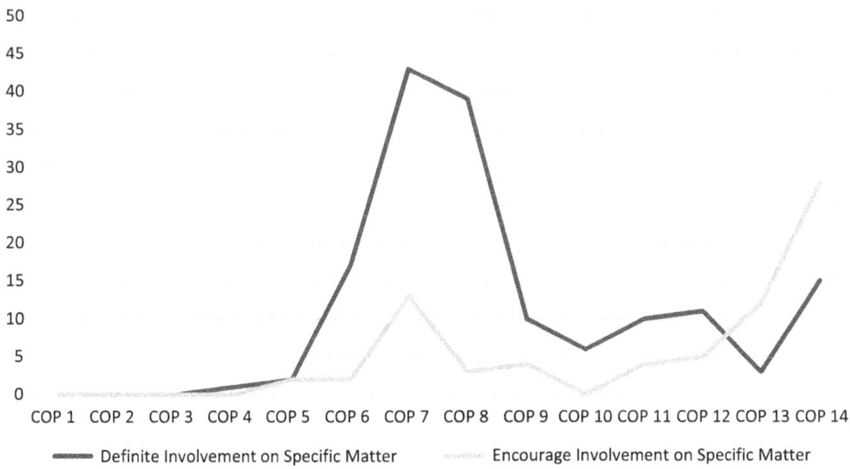

Figure 5.4 'Definite' vs. 'Encourage' Involvement on a Specific Matter Over Time.

view of a backlash or change of heart among parties after COP 7. Looking to how these two descriptions map to different cross-cutting issues at their respective high points (COPs 7 and 8 for definite involvement and COPs 13 and 14 for encourage involvement) provides further insight. More than half of all the texts assigned the directive version of the frame fall within the traditional knowledge cross-cutting issue during COPs 7 and 8. Texts assigned the less directive description that encourages participation are more spread out in COPs 13 and 14, with half of the texts spread across 5 different cross-cutting issues, including traditional knowledge. This reveals that directive language about participation was introduced to a great extent within the cross-cutting issue that is the 'natural home' for local community voices. Arguably, that gain was followed by some backlash where CBD parties shied away from such language, but a less directive version of the description emerged in later COPs among a broader range of cross-cutting issues. This suggests that using the small discursive spaces afforded within the CBD can eventually lead to the creation of spaces for participation in other areas.

Another pairing of participation descriptions that use more and less directive language also appears among the most frequent in the analysis, supplying a useful point of comparison. They concern who has power or control over participation processes, thus introducing an element of accountability. Figure 5.5 shows the frequencies over time for the 'Named actor should ensure participation' and 'named actor should support participation' frames.[3]

These descriptions or frames are less frequent in terms of overall numbers compared to those discussed earlier, but they are more stable over time.

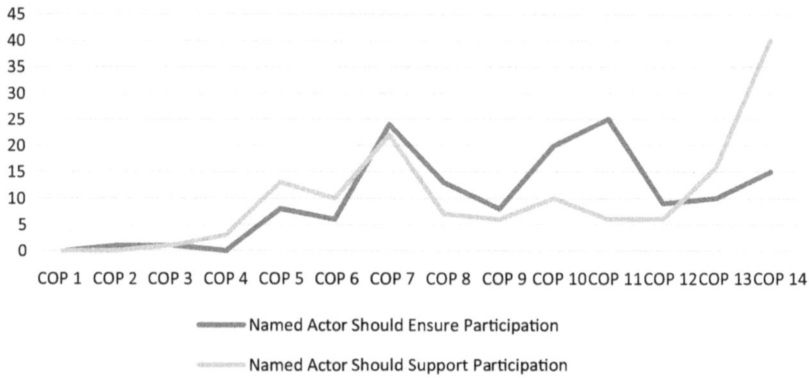

Figure 5.5 Named Actor Should 'Ensure' vs. 'Support' Participation Over Time.

This is an encouraging finding – naming actors responsible for ensuring or supporting local community participation implies accountability, a fundamental element of democratic involvement, and the relative stability of this description over time suggests a degree of acceptance of this fact. Regarding the levels of participation, the results support the view suggested earlier. The more directive description, where the named actor should ensure participation for local communities, places participation at the international level for 27 percent of texts, and the domestic level for 39 percent of texts. The less directive version, on the other hand, refers overwhelmingly to the domestic level, assigned to 71 percent of the texts. This supports the reading of a backlash against the use of directive language about local community participation at the domestic level advanced earlier. It provides further evidence to confirm a trend for more directive language at the international level, and less directive language at the domestic. The correspondence of these two descriptions with cross-cutting issues underlines the same view, as the more directive frame is more common within the traditional knowledge issue, while the less directive version (supporting rather than ensuring participation) is more evenly spread.

The findings so far suggest that CBD COP decisions emphasize that participation should take place in domestic contexts, but stop short of language that suggests this is compulsory or requires domestic actors to take direct responsibility. States may interpret directive language as encroaching on their sovereignty. The evidence from the 'definite involvement' description of participation, as well as the peak at COP 7 noted in Chapter 4, support this. CBD COP decisions appear to reserve more directive language for the international level, where the sovereignty issue is less immediate, and responsibility for accountability or compulsory processes fall on the shoulders of international actors. A further finding that arises from the analysis of these two pairs of descriptions is that participation tends to be defined in more directive ways within the traditional knowledge cross-cutting issue. When descriptions move into other cross-cutting issues, less directive versions of the same descriptions tend to be used. In this view, COP decisions may (whether they intend to or not) provide an international space where local communities can participate to raise themes of local relevance. In turn, these could be used for leverage in domestic politics in a 'boomerang' strategy, underpinned by less directive but still useful language about participation (Keck and Sikkink, 1998; Robinson, 2015).

Other internationally oriented descriptions of participation that appear less often in CBD COP decisions also hint at this last possibility. The frame 'indigenous issues body' is assigned where COP decisions mention one of these bodies as an interlocutor, which is in line with a view of the CBD as an international actor willing to bear scrutiny from indigenous and local bodies. Over half of the texts that refer to funds for participation by indigenous peoples and local communities also refer to the international level, suggesting willingness to foster and guarantee participation in a concrete way. References to the UNDRIP in CBD COP decisions can also be argued to support this view. The UNDRIP is a frequently cited source when indigenous peoples claim their rights at the international level, and thus suggests a move – again, whether intended or not – that provides space for local claims for inclusion. Though the UNDRIP is only referred to 7 times in the sample used for this analysis, the change in terminology to include the formulation 'indigenous peoples' (discussed in Chapter 4) is significant. Indeed, Robinson (2015) points out that the CBD is an arena used by various indigenous peoples to promote their self-determination in national contexts.

The prognostic frames discussed so far ('definite/encourage involvement' and 'measures to improve') do not provide much information about actual mechanisms for participation. The remaining two prognostic frames among the most frequent to emerge from the analysis furnish some more detail. They are 'inputs' (calls for information to be subsequently used in decision-making) and 'implementation' (roles for indigenous peoples and local communities in carrying out the CBD's work). Together, these two descriptions suggest that local community participation in the CBD can be interpreted with reference to the idea of civic environmentalism. In this view, local community participation is seen as crucial for fair and effective environmental policies (Bäckstrand and Lövbrand, 2016; Sconfienza, 2017). Within civic environmentalism, participation is more than simply what is due to affected stakeholders: it is necessary for better ways of resolving environmental issues. This fits with the 'inputs' frame in CBD COP decisions, which highlights the role of information from indigenous peoples and local communities for decision-making, as well as the 'implementation' frame where the same groups are assigned a role in applying decisions on the ground. Together, these two frames suggest the CBD sees participation as part of good governance.

Beginning with 'inputs', the distribution of this description is more evenly spread compared to the others illustrated in Figure 5.3. Its numbers remain among the lowest until COP 8, after which point it remains quite stable, with a notable rise in COP 14 that is not, as might be expected, confined to the Rutzolijirisaxik Guidelines on the repatriation of traditional knowledge (Convention on Biological Diversity, 2018). Unlike the other frames reported in Figure 5.3, 'inputs' does not peak at COP 7. Its increase in frequency after that meeting may indicate that it is part of the response to a perceived intrusion on national sovereignty stemming from language obliging participation by indigenous peoples and local communities in domestic contexts. The levels where inputs are called for are in line with this argument. Indeed, calls for inputs emerge as one of the main paths for direct participation by indigenous peoples and local communities at the international level, as 81 percent of texts refer to it. A further 12 percent refer to a combination of the domestic and international levels. This indicates that only a small share of participation by indigenous peoples and local communities via 'inputs' is seen as subject to State actor gatekeepers.

In a broader view, this finding provides further reason to argue that opportunities for local communities to influence CBD decisions currently appear to be greater at the international level compared to the domestic and local levels following a retreat from domestic participation after COP 7. Although the decisions taken at that meeting continue to apply, they have not increased opportunities for participation in other areas to any great extent since their adoption. This view, where local communities can use the international level, and the cross-cutting issue of traditional knowledge in particular, to expand participation opportunities both throughout the different areas of work of the CBD and, eventually, in domestic contexts, is also supported by the correspondence of the 'inputs' frame with cross-cutting issues. Around half of the texts (52 of 105) fall within the traditional knowledge cross-cutting issue, and the remaining texts are spread across a range of others. In COP 14 – the high point for 'inputs' descriptions – the spread is fairly even across 8 different issues including capacity-building, new and emerging issues (which mostly concerned synthetic biology at this meeting), traditional knowledge, sustainable use, and protected areas.

Calls for 'inputs' in CBD COP decisions thus provide a clear mechanism for local community participation, often at the international level and on issues linked to traditional knowledge and customary sustainable use. A closer look at the texts reveals that most of the provisions call on these groups to submit information, case studies, views, or perspectives, and to review and comment on CBD matters and processes, in clearly defined areas. The texts often include indigenous peoples and local communities in the same breath as States, asking both to contribute in the same ways. Whether indigenous peoples and local communities are aware of and able to package information in the ways required in CBD COP decisions is a crucial question to evaluate the democratic potential of this path for participation – particularly if States provide inputs that may compete with local community views. Usually, when a CBD COP decision calls for inputs, a notification is posted on the CBD website. Groups that represent indigenous peoples and local communities, and regularly attend CBD meetings of different types, are aware of these. What these groups and other actors, including CBD bodies such as the Executive Secretariat as well as its parties, subsequently do to disseminate calls, and to encourage and support local communities to submit inputs, is unclear.

This observation points to a link with bottom-up paths of participation, discussed with reference to community protocols in the following section. Bottom-up processes may mean that local communities have already produced information that can be brought to the attention of the CBD via the 'inputs' path. Some of the less frequent descriptions of participation that also register peaks in COP 14, co-occurring with this high point in references to inputs from local communities, though not assigned to the same texts, point to the idea that CBD parties are also aware of the potential of bottom-up participation of this kind. Descriptions of indigenous peoples and local communities as 'experts' that may benefit from 'capacity-building' to bolster participation, as well as descriptions referring to the need for participation from indigenous women and youth, and support for 'participatory models', all register higher frequencies in COP 14 and point to bottom-up processes. Despite this, the question of whether communities perceive this opportunity for participation remains crucial. Scholarship on social movements in particular suggests that actual opportunities matter very little unless activists are aware of them and consider them genuine (Tarrow, 1998).

The distribution over time of the 'implementation' description is similar to that of 'named actor should support participation'. It shows a first peak in COP 7 (albeit less marked than the 'definite involvement' and 'measures to improve' participation descriptions) and a second, higher peak in COP 14. In contrast to 'inputs', which suggest a strengthened autonomous role for local communities at the international level, 'implementation' indicates a stronger role for States. Only 10 percent of the texts place participation at the international level, and among these most refer to places in expert working groups on implementation and monitoring. The local level is conspicuous by its absence (as it is throughout the analysis), though the definition applied to local participation is, as outlined in the introduction to this chapter, strictly limited to actors that are entirely separable from States. The low percentage of texts that refer to local-level participation thus suggests that implementation tends to involve sub-national State actors. Indeed, 65 percent of the texts assigned the 'implementation' description refer to the domestic level, though 9 percent combine the local and domestic levels. This lack of purely local participation is noteworthy in light of literature that suggests local spaces where indigenous peoples and local communities can shape the implementation of the CBD are growing, for example through the recognition of community protocols (e.g. Bavikatte, 2014) and ICCAs (Jonas, 2017). Nevertheless, in the view suggested earlier, such bottom-up initiatives may contribute to international participation via the inputs path. This could constitute a first step towards wider recognition of local implementation, as discussed in the following section.

The distribution of texts among cross-cutting issues at the two high points for the 'implementation' description – at COPs 7 and 14 – confirms the argument made so far. Texts are concentrated in the traditional knowledge cross-cutting issue in COP 7, and spread to others in COP 14, notably on the sustainable use of biodiversity, and decisions adopting voluntary guidance for a sustainable wild meat sector.[4] Since the implementation of COP decisions mainly takes place in domestic contexts, the scope for participation in different cross-cutting issues is tempered by the potential actions of State gatekeepers. Research on how local communities are involved in the implementation of CBD decisions in different national contexts would shed further light on this. Overall, 'inputs' and 'implementation' suggest a divide between international-level inputs from indigenous peoples and local communities on the one hand, and domestic settings that filter their participation in implementation on the other.

Concluding, the general view suggested by this brief analysis of descriptions of participation by indigenous peoples and local communities in CBD COP decisions is as follows. The most frequent description of participation ('definite involvement on a specific matter') is confined to COP 7, underlining the importance of analysis over time. This description often refers to the domestic level, and this, it is argued, appears to have sparked a backlash against potentially binding commitments for local community participation in domestic settings among the parties to the CBD, as this description is not frequent in later COPs, while the use of conditional language for domestic participation becomes more prevalent. After COP 7, more directive language (that conveys the sense of an obligation to ensure participation) tends to be reserved for participation at the international level. Less directive language also tends to be adopted when participation by indigenous peoples and local communities is mentioned beyond the cross-cutting issue of traditional knowledge. Together, this suggests that local communities can use international paths of participation to raise local themes linked to traditional knowledge in a first instance. These spaces for participation then begin to take hold, slowly and with more conditional language, in other cross-cutting issues.

Amongst the most frequent descriptions of participation, 'inputs' and 'implementation' by local communities and indigenous peoples are the most detailed mechanisms or prognostic frames. Descriptions referring to 'inputs' provide a clear path to international participation for local communities, but whether this opportunity is adequately communicated and perceived is an open question. There is some limited evidence that the parties to the CBD seek to support bottom-up preparation for local communities and indigenous peoples to use this path in COP 14, since descriptions that define them as experts and support participatory processes, capacity-building, and participation by indigenous women and youth accompany a spike in calls for input here. Implementation forms another clear opportunity for local community participation, but CBD COP decisions mostly refer to the domestic level where this is concerned, meaning that States play gatekeeping roles. Opportunities for implementation are therefore defined in national contexts.

Overall, the evidence from the analysis of participation in COP decisions also underscores the potential of so-called 'boomerang' strategies (Keck and Sikkink, 1998), where local communities can use the international level for leverage at the national level. Here, local communities may seek to participate by providing inputs to the international level or by raising themes linked to traditional knowledge. Through inputs paths, themes of local importance may be taken into account in decisions later implemented within States. Through raising traditional knowledge themes that later spread to other areas of the CBD's work, further spaces for participation can be opened at the international level of the CBD to feed into similar boomerang effects.

What might this mean for the local communities presented in Chapter 2? From a local perspective, descriptions of participation in CBD COP decisions that lack detail about actual mechanisms are not easy to actually use, with the exception of the Ad Hoc Working Group on Article 8(j). Local communities, such as those from Chapter 2, could use links with representative groups active in this body to raise themes of importance to them in this forum. In line with this and more broadly, the cross-cutting issue of traditional knowledge provides a defined thematic area where local communities might also seek to raise matters that affect them. Clearer mechanisms for participation are available via 'inputs' and 'implementation'. Many of the communities presented in Chapter 2 are involved in activities that could be imagined to play a part in implementing the CBD. The wildlife management roles of the Khwe communities in Namibia, the efforts to propagate plants outside protected areas of the Kukula in South Africa, the actions of the Kelabit community to boost cultivation of the unique Bario rice variety in Malaysia, the goal of returning to sustainable patterns of pastoralism in Greece, and the struggle to protect the delicate freshwater supplies in the salt flats around Olaroz and Salinas Grandes in Argentina: all of these can be linked to the objectives of the CBD. Linking this point to the findings in the analysis about the gatekeeper role played by State actors in domestic implementation, the case studies suggest that naming actors as responsible for participation processes is of particular importance. This is reflected in the comparison of the cases discussed in Chapter 3, and the findings about the importance of procedural guarantees in particular. To make participation in implementation effective for both communities and the objective of conserving biodiversity, accountability over participation processes could provide one way to meet the needs for procedural guarantees. The analysis shows that the CBD has moved away from more directive language about accountability over participation in domestic contexts, opting instead for more conditional language. The case studies in this book suggest instead that a more forceful approach may be needed.

The local communities discussed in Chapter 2 also hold knowledge and experience that could be useful to the CBD via the 'inputs' path. Community protocols are potentially useful tools when imagining how such 'inputs' could be expressed and eventually communicated to the international level. Among the case studies presented, those with community protocols in place had made more progress towards the aims they had set themselves (Parks, 2018). One reason for this was that the process of developing and drafting a community protocol led to the definition of aims based on certain knowledge, experiences, and ways of life. By codifying these in protocols, communities engage in a form of 'intercultural translation' for external actors (de Sousa Santos, 2014). A similar process could potentially apply for the international level where the 'inputs' path of participation in the CBD is concerned. The following section discusses this in more depth, considering how community protocols might address the international level.

The view from below – community protocols and international participation

Community protocols have the potential to function as conduits for participation at the international level, but under what circumstances, and how exactly might they be used to address the power asymmetries that clearly exist between local communities and other actors? These issues are discussed in the following pages, which also seek to place community protocols as tools within the wider 'repertoires of contention' available to local communities, and thus within broader literature on social movements and activism. Indeed, it is important to underline that community protocols are only one potential tool among others. They may be attractive options for some communities, but an array of collective action choices exist. Which option – if any – a community chooses is context-dependent, and may shift over time. Scholars of social movements suggest that collective action can be understood along a continuum ranging from less contentious action in more open national contexts, to more contentious action in more closed national contexts (McAdam, Tarrow and Tilly, 2001; Meyer, 2004). Community protocols lie at the less contentious end of this continuum as tools that aim to open up meaningful dialogues with various external actors. Where a local community has no reasonable expectation that meaningful dialogue can be achieved, other types of collective action may be chosen. Community protocols may also be one tool among others for a community, as in some of the cases presented in Chapter 2.

In addition, community protocols are first and foremost tools that are used in national and local contexts, a point also underlined in the discussion of recognition from without in Chapter 3. They aim to provide the basis for dialogue with external actors by explaining how communities take decisions, their aims and beliefs, and other matters. While the discussion here focuses on how community protocols might be used for international participation, this is understood as an additional feature. Their primary role remains domestic. Considering these two points, about the openness of domestic contexts and the primary function of community protocols in regulating relations with domestic actors, also suggests that community protocols will only be developed where they are deemed likely to contribute positively to relations between communities and relevant external actors.

These considerations also underline that local communities preoccupied with relations with actors closer to home will not necessarily be interested in international participation. This is clearly not true of all local communities: some engage with the international level on the basis of a rejection of the international system and opt for more contentious paths that challenge their very foundations (Dryzek, 2005). Others may be interested in international participation, but unaware of the specific roles that international organizations play in local contexts. Some are both aware of these roles and interested in international participation aimed at influencing international organizations however, even where they maintain healthy levels of skepticism. This is evidenced in a growing body of scholarship about the role of indigenous peoples' representatives and their work in various international negotiations (Brand and Vadrot, 2013; Marion Suiseeya, 2014; Witter *et al.*, 2015; Eastwood, 2019; Marion Suiseeya and Zanotti, 2019). Bearing all of this in mind, a discussion of community protocols as one tool in the repertoire available to local communities remains interesting, particularly given the findings of the analysis in the previous section and in view of the volume of literature devoted to more contentious environmental action compared to less contentious strategies (e.g. Rootes, 2004; Dietz and Garrelts, 2014).[5]

Community protocols are 'documents developed by local communities to record, among other things, information on their practices, institutions, customary law, rights and visions of development' (Parks, 2018: 87–88). At the level of the CBD, community protocols were explicitly recognized in the Nagoya Protocol (Convention on Biological Diversity, 2010) following the efforts of representatives of indigenous peoples and local communities – another reason to reflect on their potential for international participation. Inspired by work by, among others, Darrell Posey (2004), community protocols are legal tools that can help defend the rights of communities by linking international, national, and customary law and policy. They are drafted following processes where communities codify existing practices and underline their rights, as well as outline future visions (Jonas, Bavikatte and Shrumm, 2010). Bavikatte, Robinson and Oliva (2015: 20), define a community protocol as:

> a convivial legal tool that is collectively developed by a community. It is aimed at those who want to engage the community and it seeks to

> articulate the community's way of life, history, customs, and decision-making processes. It begins a dialogue that goes beyond a purely instrumentalist or use-value interaction and embarks on building a relationship. Through its community protocol, a community says to the listener: if you want to access our lands, biodiversity, and knowledge, then you need to hear our story, you need to understand what these things mean to us, what our values are, and how we make decisions.

In this definition, community protocols are used to begin dialogues with actors that go to communities seeking access to some goods the community holds. This highlights community protocols as defensive tools. Yet they are also described as proactive tools where they outline communities' visions for future development (Jonas, Bavikatte and Shrumm, 2010). It is in this more proactive view that community protocols can be understood as tools for international participation. Tobin (2013) discusses community protocols and benefit-sharing from the perspective of legal pluralism, noting that the 'adoption of such protocols would have the potential to significantly influence the design of national, regional and international law and policy' (159). In this view, community protocols can feed information to multiple levels, including the international. This is in line with the analysis in the previous section, where a path for local participation at the international level was identified within CBD COP decisions via calls for 'inputs'.

Community protocols could thus be used as ready-made sources for local communities to pass information to the international level. How exactly this might take place requires more attention to be paid to both practical issues and to how these documents act within social relations of power. How can communities perceive and exploit opportunities for international participation? And how exactly might community protocols convey what Bakivatte, Robinson and Oliva (2015) call a community's 'space within' to become 'convivial legal tools' for participation at the international level? The following reflections turn first to the question of how community protocols might address power relations, and second to related practical questions, drawing on examples from the case studies.

To function in social relations of power and exercise what Fung and Wright (2003) term 'countervailing power', community protocols must translate the worldviews of communities. As de Sousa Santos (2014) argues, existing institutions often view local knowledge as superstition and thus discount it, necessitating the kind of 'intercultural translation' that community protocols can provide. One of the aims of Chapter 4 was precisely to identify the extent of discursive spaces available for local communities in the CBD. It found that there was a small but stable space for local voices that community protocols could contribute to. To understand this in more detail, a discussion of how community protocols challenge the discursive basis of the CBD as viewed by the critical literature ties in with the analyses presented in Chapter 4 and the first part of this chapter. It will show where community protocols can address power imbalances by challenging dominant discourses, and thus how community protocols can be viewed as tools for growing the small discursive space for local voices in the CBD. The discourses that underpin global environmental governance according to the critical literature reviewed in Chapter 1 (Local communities and global environmental governance) are the nature/culture divide and capitalism/colonialism. One major consequence of these discourses is the external imposition of policy decisions and the exclusion of local communities. The latter provides the key for discussing practical questions about how community protocols may be used for international participation.

Capitalist and colonial discourses are linked in the literature. Colonial conquests are seen as one particularly damaging strategy for capitalist gain, which continues to shape politics at all levels in the present day. Literature on global environmental governance makes the same claim (e.g. Natarajan and Khoday, 2014), while work on access and benefit-sharing in particular underlines how this policy reproduces patterns of exploitation of resources in the global South for the benefit of the global North. The recognition of community protocols in the Nagoya Protocol is referred to when claiming their potential for addressing this discourse (Bavikatte, 2014), yet this recognition can be interpreted in two ways. It can be seen either as a victory, creating spaces for non-dominant voices or, conversely, as an attempt to co-opt local voices into the colonialist north/south relations attributed to benefit-sharing at the international level. The potential of community protocols to address capitalist discourses at the international level relies in this view on their power to demonstrate the range of ways in which local communities define benefit beyond (though not necessarily excluding) monetary benefits. The common themes of concern that emerged from the case study comparison presented in Chapter 3 thus feed into an assessment of the potential of community protocols. The concerns of local communities can also be their strengths. Benefits, at least in the Nagoya Protocol, are not precisely defined (Morgera, 2016), demonstrating that the range of goods that local communities understand as benefits can challenge capitalist definitions at the international level. Whether expressing this is possible for local communities is tied to more practical questions discussed later in this section.

Another potential limitation to challenging the colonialist/capitalist discourse at the international level is the CBD's recognition, in line with a predominant ideal in international law more generally, of State sovereignty over natural resources (Reimerson, 2013).[6] This highlights that while community protocols can address issues dealt with in global arenas, they must primarily address a range of national actors to challenge dominant discourses. The participatory potential of community protocols necessarily regards both domestic and international contexts in this view, as suggested earlier. Local community concerns are domestic, focused on actors close to them, and this is underlined by the recognition of sovereignty over natural resources in international law. The potential for community protocols in this regard thus varies between States as well as peoples or communities, again as suggested earlier. One clear factor that plays a role in determining how community protocols might fare as conduits for international participation, based on an assumption that colonialist/capitalist discourses are also dominant in domestic contexts, is that a key element determining the likelihood that State actors will consider local community claims is the economic worth of the natural resource in question (Nelson, 2010). The local case studies presented here illustrate this point. In Argentina, for example, lithium was named as a 'strategic resource', which was seen as a factor in the perceived failure to follow rules on impact assessments (see Chapter 2: The 33 communities: Salinas Grandes and Laguna de Guayatayoc). In the Greek case, local pastoralists in Ikaria felt their access to grazing land was under threat in part due to the economic needs of the central government (see Chapter 2: Traditional pastoralists in Ikaria).

How far community protocols might address capitalist/colonialist discourse in domestic contexts thus depends on a number of factors, including the economic worth of the natural resource and the openness of the legal and political landscapes to local community claims (more open political and legal regimes will furnish more access points or opportunities to be heard). The need to challenge dominant discourses at different levels, discussed in the previous section as part of a 'boomerang' strategy (Keck and Sikkink, 1998), is underlined by these considerations, as is the link between the international and domestic contexts. For the specific case of community protocols, their potential success in making local voices heard in domestic contexts is inextricably linked to their potential for international participation. If they are able to convince actors in domestic contexts, where the principle of sovereignty over natural resources holds, domestic gatekeepers could reasonably be expected to either pass their experiences to the international level, to encourage local communities to do so directly, or both, potentially multiplying inputs originating at the local level. The South African case hints at how this might be possible. Members of the Kukula, a group with the longest standing community protocol amongst those studied in this book that made noteworthy progress in relations with a range of external actors, were invited to participate in the Steering Committee of the South African National Biodiversity Institute (see Chapter 2: The Kukula Traditional Health Practitioners Association). This illustrates how community protocols may contribute to participation in national contexts that can conceivably translate in turn into international participation, since this is one type of group that the CBD seeks inputs from. This could take place in addition to direct participation in negotiation processes, as well as participation via community protocols brought to the attention of international actors by NGOs (see below).

Conversely, where community protocols are not suited to opening meaningful dialogues with domestic actors due to closed domestic opportunity structures (Tarrow, 1998; Meyer, 2004), local communities may opt for other, more contentious strategies. A contentious approach is in turn more likely to characterize local community approaches to any international level participation, channeling efforts within what Dryzek (2005) labels as 'green radical' discourses that challenge the system of global environmental governance rather than participating in its reform from within. The existing literature on social movements suggests this correlation exists between contentious approaches at different territorial levels, where more contentious approaches to participation in local and national contexts also shape approaches to international processes (e.g. della Porta *et al.*, 2006). Yet studies of contentious politics among indigenous peoples in particular show that even where domestic contexts are closed, more conventional strategies can be followed at the international level. A good example here is the activism of indigenous groups for the UNDRIP (Cittadino, 2019). This suggests that even where community protocols do not have clear impacts in opening dialogues within national contexts, the international path may still be open, allowing these tools to give voice to local communities. This may then eventually have knock-on effects in domestic contexts as advanced in the 'boomerang' model (Keck and Sikkink, 1998) and evidenced by the ongoing influence of the UNDRIP (Cittadino, 2019).

The 'nature/culture divide' is a second major discourse attributed to global environmental governance by the critical literature. This is a view that underpins the capitalist/colonialist discourse: expansion and economic growth are seen as central to the latter, and this is based in turn on a dichotomous understanding of 'nature' as separate from humankind. Humankind dominates nature, using it as a set of resources for economic production (see e.g. Castree, 2008, for a detailed discussion). This line of argument is also extended, according to many scholars, to indigenous peoples, who are grouped in with 'nature', allowing the defense of colonial conquest as somehow 'civilizing' (Said, 1978; on the CBD Reimerson, 2013). Since dominant discourses displace others, a consequence of the centrality of this anthropocentric and dichotomous view is the exclusion of more holistic discourses or worldviews. This is attributed in the literature to green economy approaches, including benefit-sharing and payments for ecosystems services (Natarajan and Khoday, 2014; Wilkinson, 2014). Two possible interpretations can be applied to such policies: either they can be seen as striving to and capable of accommodating other worldviews through their attempts to redefine 'value' in ways that move beyond an economic focus (Morgera and Savaresi, 2013; Sikor, 2014), or they cannot, and work to reinforce capitalist/colonialist approaches. The discourse analysis of CBD COP decisions presented in Chapter 4 uncovered a small but qualitatively evolving recognition of non-capitalist worldviews, though little evidence that such worldviews were placed on the same level as more monetary approaches.

These contrasting interpretations represent different scenarios for community protocols and their potential for challenging the discourse of a nature/culture divide. The more pessimistic scenario leaves little room for maneuver, yet is challenged, in a limited way, by the findings of the discourse analysis presented in Chapter 4 that suggests a discursive 'chink in the armor' of the nature/culture divide. While acknowledging that opportunities are narrow, there is therefore reason to consider how community protocols might widen these via international participation. Once again, looking to the case studies presented in Chapter 2, the chances for international influence on the subject of respect for other worldviews seems to be conditioned by domestic experiences. The Namibian case is a good example. One reason the Khwe community within Bwabwata National Park decided to pursue a community protocol process was concern over cultural reproduction. They felt that the traditional skills used in their roles as participants in park management were at risk (see Chapter 2: The protocol of the Khwe community of Bwabwata National Park). This was also the case to some extent for the South Africa case, where a common pool of traditional knowledge was established to protect it for future generations (see Chapter 2: The Kukula Traditional Health Practitioners Association). In this view, the domestic success of a community protocol in terms of prolonging local community knowledge that may include systems of belief, institutions, and accompanying ways of life has effects on the chances that a community protocol can communicate such ideas via international participation. This is supported by findings that underline how communities that have suffered long histories of marginalization experience the destabilization of worldviews over time (Kaldor, 2003; Bayrak, Tu and Burgers, 2013, for a global environmental policy example).

The worth of community protocols as expressions and recordings of alternative worldviews is thus clear both in the perspective of cultural reproduction and in the perspective of international participation. Expression of a worldview may form the basis for dialogue between dominant and minority worldviews (though not automatically), and may also allow a community to rediscover or (re)define worldviews at risk of disappearing. Long-standing marginalization may be a factor in the latter, including structural inequalities suffered by community members (Delgado, 2016) – though this has also been identified as a source for the galvanization of struggle in some groups (Almeida, 2003). The Argentina case, for example, involves a community with a strong worldview transmitted through a history of struggle. By recording their worldview in a community protocol, they seek to communicate their reasoning to external actors. Entitled 'tracks in the salt', where 'a track represents a path to follow', this community protocol 'hopes to serve as a track, as a community conduit to channel our rights' (Comunidades Indígenas de las Salinas Grandes y Laguna de Guayatayoc de Jujuy y Salta, 2015: 5) (see Chapter 2: The 33 communities: Salinas Grandes and Laguna de Guayatayoc). The South Africa case shows that communities of interest can also elaborate elements of a worldview, as the protocol communicates the links between spiritual beliefs, customary law, and the sustainable ways in which the practitioners harvest plants (Traditional Health Practitioners of Bushbuckridge, 2010: 2; see also Chapter 2: The Kukula Traditional Health Practitioners Association). The cases where no community protocols were in place also support this – the Greek and Malaysian cases both involved some frustration amongst some community members, since they saw others as misunderstanding their worldviews (see Chapter 2: Traditional pastoralists in Ikaria; also Bario's tripartite agricultural deal). This suggests that community protocols have potential (again, as one option amongst others) to aid in attempts to communicate worldviews.

This observation links to more practical points. If community protocols are to express shared worldviews to take advantage of narrow opportunities at the international level, attention to how they are drafted is a necessary part of the equation. This theme was touched upon in the discussion of 'recognition from within' as well as organizational matters in Chapter 3, where it was noted that communities that are able to address questions about their definition and organizational forms in ways that include all members and allow them to reach consensus about wishes, institutions, aims, and other matters, were better equipped to act collectively. These communities were also those that had developed community protocols, highlighting that the process of negotiating a community protocol can allow these questions to be broached. In this view, the processes of community protocol negotiation (or similar processes linked to other types of action) are essential to meaningful challenges to dominant discourses – if a protocol does not reflect a consensus built and shared by a whole community, it will not hold real meaning for that community and thus mount no real challenge to dominant discourse.

The potential for community protocols as conduits for international participation thus relies not only on domestic contexts but also on local community processes. The case studies presented in Chapter 2 and compared in Chapter 3 provide some examples. In the South African case, the members of the Kukula Association joined together to draft a community protocol on the basis of shared interests, but members span different ethnic and language groups. Because the group had to be built from scratch as a community, particular attention was paid to consensus-building, which was credited for the perceived success of their community protocol (Sibuye *et al.*, 2012; see Chapter 2: The Kukula Traditional Health Practitioners Association; and Chapter 3: Community organization). Similarly, the 33 communities from the Salinas Grandes area in Argentina also paid attention to achieving consensus:

> In our community life we are used to carrying out different tasks in the form of "minka", work shared between us all, putting together the efforts of all of us for the benefit of the community or of one of

> its members. So we drew up this document in this way, in a communal and shared way, finding consensus between all the players in our communities, discussing each concept and each word.
>
> (Comunidades Indígenas de las Salinas Grandes y Laguna de Guayatayoc de Jujuy y Salta, 2015: 5)

In these cases, negotiating a community protocol provided a focus to concentrate efforts on questions of 'who the community is', and how it functions, in inclusive ways. Other processes may provide similar opportunities. Regarding the discursive underpinnings of the CBD, in this view community protocol processes have the potential to address the exclusion of local communities and the external imposition of policies by allowing communities to define themselves, act collectively, and present themselves to external actors with information about the roles they play in environmental questions.

Continuing to address more practical themes, once a local community has developed a community protocol, how exactly can this be used for international participation? In the first part of this chapter, the 'inputs' path was identified as one possible avenue. Based on the case studies, as well as literature on civil society, it seems that NGOs that work to support local communities are often well placed to take up the contents of community protocols and deliver them to international actors in what can be described as a 'conveyor belt' model (Kaldor, 2003). This brings another layer of complexity to the discussion of the delicate role NGOs must play to support local communities without taking over community protocol processes (see Chapter 3: Support from non-state actors as brokers). In this view, NGOs may also have a crucial role to play in bridging the gap between local communities and international actors. This is suggested in the discussion of 'inputs' earlier, where it was noted that calls for information from international actors in the CBD are placed on its website, and that representatives of indigenous peoples and local communities are aware of these. In addition, community protocols have been developed by the work of a range of actors including NGOs, and were recognized by the CBD in its Nagoya Protocol due to the efforts of NGOs among others. Finally, as noted in Chapter 3, all of the communities in the cases received support from a range of NGOs (see Support from non-state actors as brokers).

The role of NGOs or other non-state actors who support local communities thus emerges as another element that may affect the potential of community protocols for international participation. Their function as brokers speaks to the issue identified earlier in the chapter about the *perception* of participation opportunities. The existence of an opportunity in this view is of little use unless visible (Tarrow, 1998), and NGOs acting as brokers for participation in international environmental decision-making emerges as one possible way of ensuring that local communities know of these opportunities. NGOs can also act to pass on communities' views and opinions, as expressed in community protocols, to the international level. The literature on civil society, and global civil society in particular, supports this view theoretically, and is discussed in more depth in the conclusions to this book (see Chapter 6: Local communities within global civil society). Briefly, there are several central elements in theories and assessments of modern global civil society, which is defined as comprising both more and less contentious actors. First, literature on both NGOs and social movements points to the importance of networked forms of organization. Social movements themselves can be understood as networks of like-minded individuals and groups including some NGOs (della Porta and Diani, 2005). Global civil society is also understood as a field where different groups and movements are linked in more or less close relationships (Kaldor, 2003). NGOs in this perspective are either networks themselves, creating a bridge between the local and international levels, or they are members of wider networks where different groups fulfill a bridging role. Networks are also an important consideration in the norm diffusion literature, which underlines how they are used to spread and change norms, discourses, and frames (Djelic, 2004) (see Chapter 1: Norm diffusion – an interdisciplinary overview).

The importance of the brokerage role potentially played by NGOs in this type of scenario means that the critical views in the literature on civil society must also be addressed. The discussion here gives a brief overview, and the theme is picked up in more depth in Chapter 6 (Local communities within global civil society). In this literature, the history of global civil society and international NGOs in particular is located in the capitalist/colonialist discourse. This view of 'neoliberal' (Kaldor, 2003) or 'NGOized' (Choudry and Kapoor, 2013) civil society highlights the progressive co-optation of international NGOs through processes of institutionalization, professionalization, depoliticization, and demobilization (ibid.). To access opportunities to influence State and global actors, NGOs institutionalize and professionalize, and export these values, which bear close resemblance to the aims of State actors and international organizations, in their local-level work. The opportunities afforded to community-based activists to work with international NGOs in turn decrease more contentious local action, which is deemed necessary to challenge dominant capitalist/colonialist discourse (ibid.). Despite this general view, the literature acknowledges that some international NGOs do not fit this template, in particular those engaged in longer-term relationships with local communities and those that were founded in or have relocated to the global South (ibid.). The NGOs in the case studies presented in Chapter 2 can be counted among these, since they all provide long-term support (as opposed to services) to communities to navigate complex legal and policy landscapes. In addition to this, none of the cases concerned communities where social movements or more contentious struggles were underway.

This chapter sought to complete the discussion about where local community themes might find spaces in international arenas begun in Chapter 4, by looking at paths for participation. In the first part of the chapter, the focus was how the CBD has defined participation by indigenous peoples and local communities over time in its COP decisions. It included reflections about the levels at which participation is envisaged to take place, and the cross-cutting issues it concerns. Among the paths for participation identified in this first section was that of 'inputs', which is the clearest and most widespread opportunity for direct local community participation in the CBD. The second part of the chapter reflected on one possible mode for exploiting this path through community protocols. The discussion centered on how community protocols may act to challenge the discursive bases of the CBD, and on more practical questions about how such participation might take place. The potential of community protocols, as one tool available to local communities, was judged as intertwined with the ways they are used in domestic contexts. Specifically, how successful community protocols prove in building relationships between local communities and relevant domestic actors was suggested to affect success at the international level. The processes adopted in the negotiation and development of community protocols were also linked to their potential to express different worldviews, since inclusive processes are crucial to their authenticity. In a more practical view, the second part of the chapter discussed the role of NGOs as brokers of international participation.

The concluding chapter will open up these reflections, all of which are closely linked to a small number of case studies and one arena of international environmental governance, to broader bodies of literature. As underlined in the introduction to the book, the significance of this investigation of local communities and international governance is broader than the policy area of benefit-sharing. The findings speak to themes central to the literature on global civil society and ideas about democratic global environmental governance.

Notes

1 This section builds on Parks and Schröder (2018), which discusses the analysis of CBD COP decisions up to COP 13. Mika Schröder developed the coding scheme for levels of participation and carried out the analysis for levels, and contributed to reflections on the results. The author carried out the inductive analysis for all COPs, and level analysis for COP 14.

2 A small number of texts (12) from the discourse analysis presented in Chapter 4 were broken down into sub-paragraphs for the present analysis. The numbers of texts in the two analyses presented in Chapters 4 and 5 are thus different (1132 vs. 1144).

3 The 'named actors' concerned vary considerably, but include government bodies, actors from particular industries (e.g. tourism), agencies (e.g. international funding and development agencies), CBD bodies (e.g. the Executive Secretariat), and local groups themselves.

4 CBD decision 14/7, Annex.

5 The discussion in this section is based on Parks (2018).

6 Though State sovereignty over natural resources is by no means entirely unrestrained. See Perrez (2000).

References

Almeida, P. D. (2003) 'Opportunity organizations and threat-induced contention: protest waves in authoritarian settings', *American Journal of Sociology*, 109(2), pp. 345–400.

Bäckstrand, K. and Lövbrand, E. (2006) 'Planting trees to mitigate climate change: contested discourses of ecological modernization, green governmentality and civic environmentalism', *Global Environmental Politics*, pp. 50–75. doi: 10.1162/glep.2006.6.1.50.

Bäckstrand, K. and Lövbrand, E. (2016) 'The road to Paris: contending climate governance discourses in the post-Copenhagen era', *Journal of Environmental Policy and Planning*, 10 March, pp. 1–19. doi: 10.1080/1523908X.2016.1150777.

Bavikatte, K. (2014) *Stewarding The Earth. Rethinking Property and the Emergence of Biocultural Rights*. Oxford: Oxford University Press.

Bavikatte, K., Robinson, D. F. and Oliva, M. J. (2015) 'Biocultural community protocols: dialogues on the space within', *IK: Other Ways of Knowing*, 1(2), pp. 1–31.

Bayrak, M. M., Tu, T. N. and Burgers, P. (2013) 'Restructuring space in the name of development: the sociocultural impact of the Forest Land Allocation Program on the indigenous Co Tu people in Central Vietnam', *Journal of Political Ecology*, 20(1), pp. 37–52.

Benford, R. D. and Snow, D. A. (2000) 'Framing processes and social movements: an overview and assessment', *Annual Review of Sociology*, 26, pp. 611–639.

Bosi, L., Giugni, M. and Uba, K. (eds) (2016) *The Consequences of Social Movements*. Cambridge, UK: Cambridge University Press.

Bowman, M. (2010) 'Environmental Protection and the Concept of Common Concern of Mankind', in Malgosia, F., Ong, D. M., and Merkouris, P. (eds) *Research Handbook on International Environmental Law*. Cheltenham: Edward Elgar, pp. 493–518.

Brand, U. and Vadrot, A. B. M. (2013) 'Epistemic selectivities and the valorisation of nature: the cases of the Nagoya Protocol and the intergovernmental science-policy platform of the biodiversity and ecosystem services (IPBES)', *Law, Environment and Development Journal*, 9(2), pp. 202–220.

Castree, N. (2008) 'Neoliberalising nature: the logics of deregulation and reregulation', *Environment and Planning A: Economy and Space*, 40(1), pp. 131–152.

Child, B. and Barnes, G. (2010) 'The conceptual evolution and practice of community-based natural resource management in southern Africa: past, present and future', *Environmental Conservation*, 37(3), pp. 283–295.

Choudry, A. and Kapoor, D. (2013) *NGOization: Complicity, Contradictions and Prospects*. London: Zed Books.

Cittadino, F. (2019) *Incorporating Indigenous Rights in the International Regime on Biodiversity Protection. Access, Benefit-sharing and Conservation in Indigenous Lands*. Leiden: Brill.

Comunidades Indígenas de las Salinas Grandes y Laguna de Guayatayoc de Jujuy y Salta (2015) *Kachi Yupi/Huellas de la Sal: Procedimiento de Consulta y Consentimiento Previo, Libre e Informado para las Comunidadades Indígenas de la Cuenca de Salinas Grandes y Laguna de Guayatayoc*. Available at: https://cl.boell.org/sites/default/files/protocolo_final.pdf.

Convention on Biological Diversity (2004) *Akwé: Kon Voluntary Guidelines for the Conduct of Cultural, Environmental and Social Impact Assessment Regarding*

Developments Proposed to Take Place on, or which are Likely to Impact on, Sacred Sites and on Lands and Waters traditionally Occupied or Used by Indigenous and Local Communities. CBD Decision VII/16, Annex, Montreal.

Convention on Biological Diversity (2010) *The Nagoya Protocol on Access to Genetic Resources and the Fair and Equitable Sharing of Benefits Arising from their Utilization to the Convention on Biological Diversity.* CBD Decision X/1, Annex, Montreal.

Convention on Biological Diversity (2018) *The Rutzolijirisaxik Voluntary Guidelines for the Repatriation of Traditional Knowledge of Indigenous Peoples and Local Communities Relevant for the Conservation and Sustainable Use of Biological Diversity.* CBD Decision XIV/6, Annex, Montreal.

Delgado, N. A. (2016) 'Community protocols as tools for resisting exclusion in global environmental governance', *Revista de Administração de Empresas,* 56(4), pp. 395–410.

Dietz, M. and Garrelts, H. (2014) *Routledge Handbook of Climate Change Movements.* London: Routledge.

Djelic, M.-L. (2004) 'Social networks and country-to-country transfer: dense and weak ties in the diffusion of knowledge', *Socio-Economic Review,* 2(3), pp. 341–370.

Dryzek, J. S. (2005) *The Politics of the Earth. Environmental Discourses.* Second edition. Oxford: Oxford University Press.

Dryzek, J. S. and Stevenson, H. (2011) 'Global democracy and earth system governance', *Ecological Economics,* 70(11), pp. 1865–1874. doi: 10.1016/j.ecolecon.2011.01.021.

Eastwood, L. E. (2019) *Negotiating the Environment. Civil Society, Globalisation and the UN.* London: Routledge.

Fung, A. and Wright, E. O. (2003) 'Countervailing Power in Empowered Participatory Governance', in Fung, A. and Wright, E. O. (eds) *Deepening Democracy. Institutional Innovations in Empowered Participatory Governance.* London: Verso Books, pp. 259–289.

Gamson, W. A. (1985) 'Goffman's legacy to political sociology', *Theory and Society,* 14(5), pp. 605–622.

Jonas, H., Bavikatte, K. and Shrumm, H. (2010) 'Community protocols and access and benefit sharing', *Asian Biotechnology and Development Review,* 12(3), pp. 49–77.

Jonas, H. C. (2017) 'The Evolution of Indigenous Peoples' and Community Conserved Territories and Areas (ICCAs) in International Biodiversity Law and Conservation Policy', in Morgera, E. and Razzaque, J. (eds) *Encyclopedia of Environmental Law: Biodiversity and Nature Protection Law.* Cheltenham: Edward Elgar, pp. 145–160.

Kaldor, M. (2003) *Global Civil Society: An Answer to War.* Cambridge, UK: Polity Press.

Keck, M. E. and Sikkink, K. (1998) *Advocacy Networks in International Politics.* Ithaca, NY: Cornell University Press.

Marion Suiseeya, K. R. (2014) 'Negotiating the Nagoya Protocol: Indigenous demands for justice', *Global Environmental Politics,* 14(3), pp. 102–124.

Marion Suiseeya, K. R. and Zanotti, L. (2019) 'Making influence visible: innovating ethnography at the Paris Climate Summit', *Global Environmental Politics,* 19(2), pp. 38–60.

McAdam, D., Tarrow, S. and Tilly, C. (2001) *Dynamics of Contention*. Cambridge, UK: Cambridge University Press.

Meyer, D. S. (2004) 'Protest and political opportunities', *Annual Review of Sociology*, 30, pp. 124–145.

Mitrotta, E. (2019) *Decentralised International Cooperation: Enhancing Conservation and Sustainable Management of Transboundary Natural Resources*. University of Trento.

Morgera, E. (2016) 'The need for an international concept of fair and equitable benefit-sharing', *European Journal of International Law*, 27(2), pp. 353–383.

Morgera, E. (2017) *Reflections on 2016 UN Biodiversity Conference (Parts I & II): assessing the Mo'otz kuxtal guidelines on benefit-sharing from the use of traditional knowledge*, BENELEX Blog.

Morgera, E. and Savaresi, A. (2013) 'A conceptual and legal perspective on the green economy', *Review of European Community & International Environmental Law*, 22(1), pp. 14–28.

Natarajan, U. and Khoday, K. (2014) 'Locating nature: making and unmaking international law', *Leiden Journal of International Law*, 27(3), pp. 573–593. doi: 10.1017/S0922156514000211.

Nelson, F. (2010) 'Democratizing Natural Resource Governance: Searching for Institutional Change', in Nelson, F. (ed.) *Community Rights, Conservation and Contested Land: The Politics of Natural Resource Governance in Africa*. London: Earthscan, pp. 310–333.

Parks, L. (2018) 'Challenging power from the bottom up? Community protocols, benefit-sharing, and the challenge of dominant discourses', *Geoforum*, 88 pp. 87–95. doi: 10.1016/j.geoforum.2017.11.011.

Parks, L. and Schröder, M. (2018) 'What we talk about when we talk about "local" participation in international biodiversity law. The changing scope of indigenous peoples and local communities' participation under the Convention on Biological Diversity', *Participation and Conflict*, 11(3), pp. 743–785.

Perrez, F. X. (2000) *Cooperative Sovereignty. From Independence to Interdependence in the Structure of International Environmental Law*. Leiden: Brill.

della Porta, D. *et al.* (2006) *Globalization from Below. Transnational Activists and Protest Networks*. Minneapolis: University of Minnesota Press.

della Porta, D. and Diani, M. (2005) *Social Movements: An Introduction*. Second edition. Oxford: Wiley-Blackwell.

Posey, D. A. (2004) *Indigenous Knowledge and Ethics: A Darrell Posey Reader*. London: Routledge.

Reimerson, E. (2013) 'Between nature and culture: exploring space for indigenous agency in the Convention on Biological Diversity', *Environmental Politics*, 22(6), pp. 992–1009.

Robinson, D. F. (2015) *Biodiversity, Access and Benefit-Sharing. Global Case Studies*. London: Routledge.

Rootes, C. (2004) 'Environmental Movements', in Snow, D. A., Soule, S. A., and Kriesi, H. (eds) *The Blackwell Companion to Social Movements*. Oxford: Blackwell, pp. 608–640.

Said, E. W. (1978) *Orientalism*. London: Routledge.

Savaresi, A. (2018) 'Traditional knowledge and climate change: a new legal frontier?', *Journal of Human Rights and the Environment*, 9(1), pp. 32–50.

Sconfienza, U. (2017) *The politics of environmental narratives*. Tilburg University.

Sibuye, R. *et al.* (2012) *The Bushbuckridge BCP: Traditional Healers Organise for ABS in South Africa in Participatory Learning and Action, Biodiversity and Culture: Exploring Community Protocols, Rights and Consent, Participatory Learning and Action 65*. Available at: http://pubs.iied.org/pdfs/G03403.pdf.

Sikkink, K. (2005) 'Patterns of Dynamic Multilevel Governance and the Insider–Outsider Coalition', in della Porta, D. and Tarrow, S. (eds) *Transnational Protest and Global Activism*. New York: Rowman & Littlefield, pp. 151–173.

Sikor, T. (ed.) (2014) *The Justices and Injustices of Ecosystem Services*. London: Routledge.

de Sousa Santos, B. (2014) *Epistemologies of the South: Justice Against Epistemicide*. New York: Routledge.

Tarrow, S. (1998) *Power in Movement*. Cambridge, UK: Cambridge University Press.

Tobin, B. (2013) 'Bridging the Nagoya compliance gap: the fundamental role of customary law in protection of indigenous peoples' resource and knowledge rights', *Law, Environment and Development Journal*, 9(13), pp. 142–162.

Traditional Health Practitioners of Bushbuckridge (2010) *Biocultural Protocol of the Traditional Health Practitioners of Bushbuckridge*. Available at: http://community-protocols.org/wp-content/uploads/documents/South_Africa-Bushbuckridge_Biocultural_Protocol.pdf.

United Nations (1992) *Rio Declaration on Environment and Development*. A/CONF.151/26.

Wilkinson, K. (2014) 'Payment for "ecosystem services" and the "green economy": green-washing or something new?', *Journal of Human Rights and the Environment*, 5(2), pp. 168–191.

Witter, R. *et al.* (2015) 'Moments of influence in global environmental governance', *Environmental Politics*, 24(6), pp. 894–912.

6 Involving local communities in international decisions

Towards democratic global environmental governance

The chapters in this book illustrated first how a small number of local communities speak about issues linked to benefit-sharing, including what they consider important conditions for benefit-sharing to be able to take place. Questions about how local communities might be able to raise those issues at the international level were then discussed. The scope of the study is limited with regard to local community case studies, but was designed to speak to broader subjects while conserving an empirical and bottom-up approach to investigating issues around benefit-sharing. In this final chapter, the main findings emerging from the book are summarized and placed in the wider context of literature on global civil society and deliberative democracy in particular. The research presented in this book raises a series of questions about how to situate local actions for the environment linked to international decisions and global common goods within our understandings of civil society and the debate about democratic global environmental governance.

Summarizing the findings

To recap, Chapter 1 introduced the aims of the research and discussed some of the relevant literature, drawing from a number of disciplines. In particular, literature on norm diffusion, on the discourses of global environmental governance, on local agency in arenas of global environmental governance, and on local experiences of international environmental policies were discussed. This literature review pointed to a widely shared understanding of the dominant discourses of global environmental governance, but revealed different views about how local communities view benefit-sharing, whether their concerns are addressed in international arenas, and how they exercise agency in international decision-making processes. A wide range of work examines benefit-sharing and other international policies, including via case studies of applications in local communities. Equally, a wide range of work examines the different ways in which local communities work to conserve and protect a range of natural resources, and the challenges they face as well as their successes. Another branch of scholarship looks at how representatives of indigenous peoples and local communities act to try and wield influence during international environmental negotiations. This book has tried to at least begin to knit these approaches together, departing from local community perceptions to then examine discursive spaces in the CBD and considering practical paths for more direct participation.

Chapter 2, accordingly, started out from the local level. It introduced the brief stories of five different local communities as far as their discussions related to benefit-sharing were concerned, as well as sketching their complex governance backdrops. The aim of these case studies was organic – rather than visiting communities with pre-prepared questions about 'benefit-sharing', the idea was to hear what communities considering issues that could lead to eventual benefit-sharing discussed, and what they considered important in that view. The case studies are not complete pictures of the communities, as the visits were short, and no claim is made to have a deep understanding of their experiences. Rather, Chapter 2 presented 'snapshots'. The five cases were in Ikaria, Greece; Bushbuckridge, South Africa; Bwabwata National Park, Namibia; the Olaroz and Salinas Grandes salt flats, Argentina; and Bario, Malaysia.

The Greek case study concerned a community of traditional goat pastoralists from the island of Ikaria. This community was discussing issues about access to grazing land, which were in turn interlinked with a complex situation where EU rules and regulations, the financial crisis in Greece, and climate change had contributed to population and overgrazing problems. In the past, these sorts of issues had been regulated in communal ways, and pastoralists hoped to return to this approach and the recognition of costs and benefits it implied. The South African case study focused on a community of traditional health practitioners spanning diverse ethnic and language groups from different locations in the Bushbuckridge area bordering the Kruger National Park. This community formed an Association and drafted a community protocol in 2009 with support from various organizations. The group was working on a number of relationships with different local actors on the basis of their protocol, but were also working to update it in light of new issues linked to access to plants, amongst others. In Namibia, the case study concerned Khwe San communities residing in Bwabwata National Park. These communities were involved in a functioning benefit-sharing agreement via the Kyaramacan Association for hunting licenses and harvesting permits within the park, and received various economic and employment benefits. The communities decided to draft a community protocol to clarify a number of issues linked to cultural reproduction and their status as residents of the park and guardians of wildlife. They had completed a draft with support from the Namibian Ministry for Environment and Tourism as well as a number of NGOs.

In Argentina, the case focused for the main part on the work of indigenous communities resident in the area of the Salinas Grandes salt flats in the north of the country. Extractives companies hoped to begin mining for lithium on these flats, which belong to the indigenous communities, and had begun exploratory work. The communities did not believe that consultation processes were carried out correctly during environmental impact assessments prior to this exploratory work, and had taken various actions to highlight this as well as other issues. One action was to draft a community protocol. This documents how free, prior, and informed consent should be sought by those wishing to access the communities' land for different purposes. The community protocol had been noted by the provincial government, which suspended mining permits in the area for a time. In the same region, some community members from the areas close to the Olaroz salt flats highlighted similar concerns about consultation and consent procedures for lithium mining, though in this case permits had already been issued, and not all community members were opposed to mining. In both areas a common concern was the potential impact of mining on freshwater supplies given the techniques used for lithium mining in the area and the delicacy of the local water ecosystem in one of the most arid areas on the planet.

The last of the five case studies regards a tripartite agricultural agreement between a sub-national government, a local community association, and a farming company in Malaysia. The community concerned are the Kelabit of Bario and surrounding villages in the Malaysian state of Sarawak, Borneo. The main issue investigated in the case was the farming of the unique variety Bario rice. As younger generations have moved away from the area, cultivation levels have fallen. To protect the variety and ensure food security for the local area, the Kelabit community association secured subsidies from the Sarawak State government and struck a deal with a farming company to provide agricultural services in order to boost production. The project has been underway for a number of years, and rice cultivation has increased. However, this overall success did not come about without various issues arising as the project progressed, most of which are linked to farming techniques and, ultimately, intra-community consultation.

The results of the comparison of these five community stories was then presented in Chapter 3. The comparison, following the parameters suggested by Ribot and Peluso (2003), yielded a number of common themes. These themes were not all directly concerned with benefit-sharing alone. In line with the open and bottom-up approach to the case studies, the themes that emerged were defined by the communities themselves. As discussed in Chapter 1, the aim was to see how communities framed benefit-sharing, before investigating the extent to which these local community frames are reflected in the CBD. This stops short of a study of norm diffusion, since the analysis of mechanisms for diffusion are limited to reflections about paths for international participation for local communities. It does however provide a detailed view of the spaces for local frames in the international space of the CBD. Many of the common themes thus refer to what might be described as the 'pre-conditions' for eventual benefit-sharing, and about the importance of taking local specificities into account. Others were more directly concerned with how benefits and the processes of benefit-sharing were perceived in these communities.

A central theme, recognition, emerged from the comparison. Chapter 3 split this into two aspects, recognition from without, and recognition from within. The former regards the wide range of external actors that communities sought recognition from. Recognition itself ranged from formal to more informal types, for example recognition of a specific role fulfilled by a community (such as recognition from the Ministry of Health, which allows traditional health practitioners to refer patients in South Africa), to recognition of a community's worldview or way of life, as well as the complexity of communities and their decision-making processes (such as that explained in the community protocol of the communities of Salinas Grandes). Recognition from within, on the other hand, concerned intra-community dynamics, for example around who was included in decision-making processes. This emerged as an important factor in seeking recognition from without, since inclusive decision-making is linked to more widely shared positions in communities and thus to effective communication with external actors.

The other common themes were all linked to this central point about recognition. First, all of the cases involved support for communities from non-state actors (often, but not only, NGOs) to some degree. This raised the question of how these actors shape community processes in intentional and unintentional ways, and the question of communities becoming dependent on non-state actors' support. A discussion of short-term project funding trends and the ways in which the non-state actors involved in the case studies work underlined that the latter are committed to long-term involvement and empowerment rather than service provision. Second, how communities organized was another important theme to arise from the comparison. The case studies appear to indicate that the more hierarchically organized the community, the better the levels of access and progress made towards the goals communities defined. There were clear knock-on effects where more hierarchical organizational forms led to easier access to various procedures and actors. This suggests that communities may feel obliged to adopt such organizational forms, thus recreating a model that is convenient for State and private sector actors who seek clear individual points of contact. But these organizational forms may not sit well with the preferences or traditions of a given community. The need for the recognition and acceptance of different types of organization was discussed in this regard, though organization of some form was identified as an important component allowing collective action. Overall, the themes of support and organization were considered to be pre-conditions for communities to begin pursuing their aims – whether linked directly to benefit-sharing or to other related issues.

The next common theme was complex governance contexts. In itself, the observation that communities exist within complex contexts is not of course linked to pre-conditions to benefit-sharing. Indeed, as discussed in Chapter 1, this is one of the assumptions of the research design. Rather, this common theme intersects with another about procedural guarantees. Communities face certain difficulties as a result of the fact that they exist within ever more complex governance contexts. This is linked in turn to the central theme of recognition in the sense that this complexity obliges them to seek recognition from a wide range of actors. The discussion also focused on differences between formal models of governance and the lived realities communities faced, which were often characterized by high uncertainty about which authorities were ultimately responsible for which tasks, as well as the uneven implementation of the law.

Complexity and uncertainty are linked, as mentioned, to the common theme about procedural guarantees. This was discussed in particular with reference to information and voice for local communities. All the cases pointed to a need for information – not simply about which actors were responsible for what, but also about external actors' intentions and projects, and scientific information about environmental and other impacts. Above all, the cases pointed to the need for information from neutral parties, and for information presented in formats specified by communities themselves in a timely fashion. Voice also emerged as an important part of this equation. Specifically, communities underlined the need for clear avenues for opinions and needs to be expressed, both within communities and with reference to outside actors. Here either the absence of forums or the existence of too many forums was found to be problematic. Procedural guarantees thus emerge as another pre-condition for eventual benefit-sharing.

The final two common themes discussed in Chapter 3 refer directly to how communities frame benefits and benefit-sharing. All of the communities in the case studies underlined, in different ways, the importance of non-monetary benefits for their aims. This contrasted with the rather more monetary view of 'benefit' or 'development' that they attributed to external actors, and communities often felt that these actors misunderstood community needs falling outside these monetary terms. Communities did seek and underline monetary needs, but other benefits were also viewed as important. Once again the link to recognition is clear – communities wished to be involved in meaningful dialogues where their framings of 'benefit' could be communicated and understood by relevant external actors. This framing of benefit thus points to a further pre-condition for eventual benefit-sharing as far as the communities in the case studies are concerned: meaningful dialogue based on recognition. Dialogue also characterizes the final point to emerge from the comparison – that benefit-sharing, if and when it takes place, should be regarded by all those involved as a process rather than an event or a contract to be negotiated then simply applied. Benefit-sharing in this view is not the outcome of a limited consultation, but an ongoing process of dialogue geared towards mutual understanding. Both benefits and the way they are shared will necessarily change over time as communities' needs evolve, and a view of benefit-sharing as a process is more consonant with this.

The findings on common issues are based on a limited number of cases, which means they cannot be argued to apply to all local communities. Nevertheless, because the case selection was designed to cover local communities in different contexts, with concerns over different types of natural resources, and at different stages of discussions linked to questions of benefit-sharing, there is some basis to argue that they could well be shared by other local communities. In addition, the case comparison sought out commonalities at a more general level to seek more widely applicable themes. The common themes are also echoed in some existing literature on local communities, benefit-sharing, and natural resource management (for example Swemmer *et al.* 2015; Tysiachniouk and Olimpieva, 2019; Wynberg and Hauck, 2014). Together, these elements make the case for pursuing the analysis at the international level to see where these local community themes find discursive space. A discourse analysis of the CBD carried out with this aim in mind is nevertheless of interest in that it reveals the shifting and nuanced discursive underpinnings of this arena of global environmental governance.

Chapter 4 thus picks up the common themes arising from the case study comparison and asks whether they find discursive spaces at the international level. As discussed in Chapter 1, the constructivist basis of the research leads to a focus on meaning – the framing of benefit-sharing in local communities, and the space for these frames in the discursive landscape of global environmental governance. This motivates the choice to carry out discourse analysis. The focus is the CBD, argued by some scholars to be more open to local communities and indigenous peoples (which it considers as one group in its terminology) compared to other arenas of global environmental governance (Reimerson, 2013; Bavikatte, 2014; Affolder, 2017; Schabus, 2017). The analysis adopts discursive categories that are derived from the critical literature on global environmental governance, as well as work that is more optimistic about its potential as a host for local voices, summarized in Chapter 1. Chapter 4 provides a detailed picture of the way discourses change over time in the CBD, as well as the spaces available for the expression of the local community issues outlined in Chapter 3. The discourse analysis covers all of the CBD's COP decisions up to and including the fourteenth meeting held in 2018. It shows that there is a specific – if limited – space for local community issues in the CBD.

In more detail, Chapter 4 shows that the discursive spaces available for local community themes seems to stabilize after COP 7, held in 2004. This meeting appears as a watershed moment for the creation of discursive spaces for local community themes. It saw the adoption of the Akwé: Kon Guidelines (Convention on Biological Diversity, 2004b) and Addis Ababa Principles and Guidelines (Convention on Biological Diversity, 2004a), and a peak in discourses that can be considered as positively creating discursive spaces for local issues, specifically through talk about recognition and participation. Although Chapter 4 suggests that State parties to the CBD subsequently took a step back from these, the decisions of COP 7 remain in force and appear to have set the foundations for the general acceptance of local community recognition and participation. These discourse categories emerged strongly in the analysis, and can be linked to the issues of importance highlighted in Chapter 3. Nevertheless, the spaces for local community issues are limited. Much of the talk relevant to local community themes falls within the cross-cutting issue of traditional knowledge, rather than being evenly spread throughout the CBD's decisions. This suggests it is 'ghettoized' to some extent (Parks, 2018). In addition, the two discourse categories that are arguably the most meaningful for creating spaces for local community issues – on internal initiatives (such as community protocols) and the valorization of worldviews not based in market logics – are comparatively rare. Finally, the design of the analysis does not provide enough information about the issue of local community exclusion to draw definitive conclusions. Though preliminary research suggests that there is indeed a specific and generally positive discourse about indigenous peoples and local communities in CBD COP decisions, the location of much of this within a single cross-cutting issue continues to suggest potential exclusion from other areas of the CBD's work.

Chapter 5 takes a practical approach to investigate the possible paths for international participation – and thus for the diffusion of local framings of benefit-sharing – available to local communities. The reasoning for this step in the research is twofold. First, although the findings about the spaces for local issues in the CBD are limited, spaces are there, and warrant a closer look. Second, the literature on community protocols as well as the wider literature on the participation of indigenous peoples and local communities in arenas of global environmental governance suggests that efforts to influence the international level can open up new paths for bottom-up participation. These could conceivably create a virtuous circle of local community participation over time. Chapter 5 thus investigates how the CBD describes participation by local communities, and how community protocols – recognized by the CBD as a result of work by representatives of indigenous peoples and local communities and other like-minded actors – could be used as paths to international participation.

To investigate how the CBD frames participation, Chapter 5 thus builds on one of the clear findings of Chapter 4 – namely that talk about participation forms an important part of the CBD's decisions relevant to indigenous peoples and local communities. Participation in the international arena is an obvious way for local communities to get involved in international decision-making and to raise issues important to them. Yet the real meaning attached to participation is not revealed by the discourse analysis presented in Chapter 4. Further analysis of the texts from CBD COP decisions that refer to the participation of indigenous peoples and local communities is presented in Chapter 5, providing a view of the ways in which participation is framed by State parties to the CBD and how these framings change over time.

This analysis provides a picture of the most common descriptions of the participation of local communities in CBD COP decisions, as well as information about how these descriptions correspond to different cross-cutting issues as well as the levels (local, domestic, or international) at which participation is foreseen to take place. The most frequent descriptions of participation were found to be limited to a great extent to texts in COP 7, which texts also refer to participation by local communities in domestic contexts more often compared to other meetings. This is a potential explanation for the notable drop in talk about the participation of local communities that followed this meeting. After COP 7, more directive language suggesting that participation by local communities must be ensured, and in some cases that a particular actor should be responsible for this, refers more often to participation at the international level, and demonstrates a more even spread beyond the cross-cutting issue of traditional knowledge. This evolution suggests a retreat from talk about local community participation in domestic contexts after the 'watershed' of COP 7 in 2004, followed by a move to less directive language for domestic participation, but a wider spread to other areas of the CBD's work beyond traditional knowledge, and a greater emphasis on international participation.

Two of the most common descriptions of participation also suggest clear and practical information about how local community involvement can occur in the CBD today via 'inputs' and 'implementation'. 'Inputs', which are calls for local communities to provide information and accounts of their experiences, are particularly clear paths for raising issues of local importance at the international level, and many of these calls refer directly to the international level, meaning that communities can theoretically bypass State gatekeepers. However, whether these opportunities are adequately communicated and perceived by local communities remains an open question for further research, and includes questions about how local communities can produce the sort of information requested by the CBD. A role in the implementation of the CBD forms the second clear opportunity for local community participation, though COP decisions mostly refer to the domestic level here, which signifies that State actors are also involved according to the coding scheme applied. Overall, Chapter 5 highlights the potential of the 'inputs' path. This speaks to the idea that local community inputs, for example via community protocols, can feed into virtuous circles where international participation creates broader opportunities over time. This can work both ways, and feed into so-called 'boomerang' strategies (Keck and Sikkink, 1998), where local communities participate at the international level to draw attention to local issues, with knock-on effects in domestic contexts (Robinson, 2015).

In this view, Chapter 5 then moves on to discuss community protocols as a potential tool for local communities to participate at the international level. Community protocols are just one option among many available to local communities, and are primarily tools for acting in domestic contexts. Nevertheless, they are chosen as a focus here for various reasons, including the fact that they were recognized by the CBD in the Nagoya Protocol following work by representatives of indigenous peoples and local communities. Community protocols could conceivably be used as vehicles to exploit the potential of the 'inputs' path identified earlier in Chapter 5, and help local communities produce responses to international calls for inputs from the CBD (and indeed other bodies). In light of this, the discussion focuses on how community protocols might challenge the discourses argued to characterize global environmental governance (thus feeding into a process creating broader spaces for local issues) and on more practical questions. The potential of community protocols is argued to be intricately linked with their capacity to aid local communities in their relations with domestic actors (their primary purpose), and with the processes leading to their adoption within communities. This last point in particular was identified as a crucial element for community protocols to fulfill their potential. Open and participatory processes determine their authenticity as the expressions of a community as a whole. In a more practical light, the roles of non-state actors, and NGOs in particular, were discussed. Given the finding in Chapter 3 about the role of non-state actors in supporting communities, they are clear candidates for acting as conduits, passing the content of community protocols to the international level where this is the wish of a community.

Summing up, the research presented here was exploratory and focused on how meanings are built in social interactions by local and international actors. No specific hypotheses were made about what local communities might talk about with reference to benefit-sharing, and a similarly open approach was sought in mapping the discourses of the CBD with reference to local-level themes. What emerges from the research presented in this book thus far is a complex picture, and in many ways moves beyond the subject of benefit-sharing to underline broader issues about participation, democracy, and representation that must be addressed if fair and equitable benefit-sharing – or indeed fair global environmental governance more generally – is to be achieved. The local communities discussed in the book all have unique stories and issues to face, yet at a more abstract level some common issue areas emerge from their experiences. These common themes often concern the conditions viewed as necessary for eventual benefit-sharing to take place in a fair and equitable way. Even those common issues more closely linked to benefit-sharing – about the importance of non-monetary benefits and the understanding of benefit-sharing as an ongoing dialogue – spoke to this idea of 'pre-conditions'. The bottom line for the local communities studied was proper recognition, respect, and responsiveness. The common themes thus point not only to conditions for fair and equitable benefit-sharing, but for the proper recognition and involvement of local communities in decision-making.

These locally identified themes find some discursive ground in the international arena of the CBD, albeit limited in many different ways. In addition, descriptions of local participation in the CBD's work suggest that these discursive bases could be widened further. There are paths available for the participation of indigenous peoples and local communities in the CBD, and these have often been forged through the work of these groups' representatives and civil society more broadly. By following up on this work, local communities could contribute to broaden these spaces (Bavikatte, 2014). A particularly clear path for international participation in this vein is via the provision of information to the CBD, and community protocols could be one way to take advantage of this.

The findings of this bottom-up research on local frames of benefit-sharing and their reflection in discourses at the international level open up the opportunity to reflect on how they fit into and can contribute to the broader scholarship on participation in international decision-making. With this in mind, the following section first revisits some relevant literature on the participation of indigenous peoples and local communities in global environmental governance, then reflects on how the findings of this book fit into this. Another subsection gives an overview of Kaldor's (2003a; 2003b) work on global civil society and Choudry and Kapoor's (2013) critical view of 'NGOization' to situate local community participation more broadly, and address the potential role of NGOs that support communities. Finally, the discussion turns to questions of participation, representation, and influence in a democratic view, and places the findings in relation to suggested models of democratic global environmental governance.

Local community participation, global civil society, and deliberative democracy

The research presented in the earlier chapters raises questions about what local community participation in international arenas looks like at a time when the issues that local communities consider important find (limited) discursive ground at the international level. These questions relate to different types of power wielded by international organizations, by State parties to these, by NGOs and other non-state actors and, of course, by local communities themselves. This section expands on literature that focuses on types of local community participation not covered in this work with a view to placing bottom-up local community participation in context. The discussion also illustrates the limitations of the research. Marion Suiseeya and Zanotti (2019) point out that the majority of studies of local community involvement in environmental governance (also via representative organizations) focus on single definitions of power rather than considering the variety of guises power assumes. This book is no exception: it has focused on discursive power, understood as the foundation upon which other types of power may be built (agenda-setting power, decision-making power, etc.).

Although the focus on participation is more practical, it concerns the paths via which discursive power might be channeled rather than power itself. For this reason, not all of the questions suggested by the findings illustrated in the discussion can be answered fully – other types of power and how it is wielded need to be taken into consideration. The discussions in this section thus also go some way to filling gaps left unexplored in the research presented in previous chapters.

Marion Suiseeya and Zanotti (2019) also argue that the emphasis in much of the literature is on States and NGOs in arenas of global environmental governance, which means that many of the ways in which local communities, and indigenous peoples in particular, exert power go unheeded (ibid.). The focus in this book has been firmly on local communities with some discussion of NGOs, and discussions about the power wielded by other actors can thus help illustrate participation in global environmental governance more fully. The literature on the CBD as well as other sites of global environmental governance provides a range of views. Studies in international environmental law highlight that some of the issue areas identified as important for local communities are discussed at length and acted upon by international organizations among others. Procedural guarantees, for example, are the subject of instruments about environmental impact assessment (see Chapter 4). These instruments have been referred to in case law where local groups dispute these processes, highlighting the link between environmental law and human rights (Morgera, 2018; Marchegiani, Morgera and Parks, 2019).

Representatives of indigenous peoples and local communities are also active within the CBD. The Ad Hoc Working Group on Article 8(j) is particularly interesting in this regard. Representatives of these groups propose decisions and text in this forum (although interventions must be sponsored by a State party delegate to be taken forward) and these are later debated in COPs and often adopted after more or less far-reaching amendments (Morgera, 2017). The literature on the politics of global environmental governance points to areas where issues of importance to local communities have been raised and championed by their representatives. This is the case of community protocols and their recognition in the CBD's Nagoya Protocol, as already mentioned. Community protocols were championed by representatives of indigenous peoples and local communities as well as others (Bavikatte, Robinson and Oliva, 2015).

Other accounts that focus on representatives of indigenous peoples and local communities and how they act during COPs report mixed findings. Brand and Vadrot (2013) look at discursive power through a focus on 'epistemic selectivities' at work during the negotiation of the Nagoya Protocol and the Intergovernmental Science-Policy Platform on Biodiversity and Ecosystem Services (IPBES) during CBD COP 10. They find that these selectivities limit what is considered as valid biodiversity science within a 'logic of valorisation' of nature. This ultimately excludes the traditional knowledge held by indigenous peoples and local communities, as this falls outside the definition of valid knowledge.[1] Focusing on the Nagoya Protocol negotiations on the basis of a collective ethnographic approach, Marion Suiseeya (2014) looks at whether discussions about the meanings of justice made space for indigenous and local communities to 'introduce new ideas, values, and norms that, over time, may shift the value orientation of global environmental governance' (103). She finds that negotiations were instead dominated by debates over instruments, meaning that while the Protocol did signify a step forward for these groups in other ways, it did not leave discursive space for negotiating fundamental meanings of justice. With regard to the representatives of indigenous peoples and local communities and their influence at COP 10 as a whole, however, the same author and her colleagues do uncover 'moments of influence' where indigenous peoples and local communities used their relations with State parties to wield informal power (Witter *et al.*, 2015).

Using the same collective ethnographic approach, Marion Suiseeya and Zanotti (2019) widen their focus to include different types of power in order to analyze how indigenous peoples were able to influence the outcomes at COP 21 of the UNFCCC, better known as the Paris Climate Summit of 2015. They argue that narrow focuses on single types of power – for example discursive power, or the more formal power exerted by State parties – can obscure the power exerted by indigenous peoples' groups. They find that indigenous peoples' groups were able to influence proceedings at COP 21 despite formidable structural barriers. This was achieved through multiple avenues, including: online work where 'virtual publics' were used to remind negotiators of their global audience of citizens (see Garrido and Halavais, 2003); 'revolving doors' where indigenous people were also State party delegates; protests; and more conventional strategies including press conferences and lobbying (Marion Suiseeya and Zanotti, 2019). Together, these actions contributed to the addition of indigenous rights in Article 2 of the Paris Agreement, though hopes about their addition to the operational part of the treaty did not come to fruition (ibid.). Eastwood (2019), who also follows an ethnographic approach to study interactions between civil society and indigenous peoples' organizations at meetings of the UNFCCC, argues that these groups find few discursive openings to overcome the fundamental 'disjuncture between the exigencies of the current global economy and the environmental problems that the UN bodies are designed to address'. This suggests there may be more scope for indigenous peoples and local communities to influence global processes through rights discourses, since these are another founding concept of the United Nations and liberal democratic States (see Morgera, 2018).

This brief look at studies concerning the actions and impacts of representatives of indigenous peoples and local communities in arenas of global environmental governance suggests that a main focus when seeking influence is Conferences of the Parties – high level intergovernmental negotiations that punctuate the ongoing, everyday work of global environmental governance every few years. Other studies of involvement in global environmental governance focus on a more everyday view; work on CBNRM and political ecology studies in particular tend to focus on how global policies affect and are used by local communities on the ground (see the discussion in Chapter 1). Work on the environmental and climate justice movements, though not reviewed in this book, tends instead to consider local communities, their representatives, and other NGOs to the extent that they form a part of wider social movements that challenge the system of global environmental governance (and of global governance systems) from the outside.

The findings presented in this book point to a complementary focus on how local communities can influence the everyday work of global environmental governance. First, the common issues of concern all revolved around the idea of recognition. This was the reason for a focus on discursive power in the analysis of COP decisions, since recognition (of different aspects of communities and their worldviews) emerged as the most basic pre-condition for any eventual benefit-sharing agreement to work in fair and equitable ways. Second, the findings of the discourse analysis suggested a small area where local communities could find space within the CBD to push issues of importance. Third, and of the clearest relevance here, the analysis of participation as described in CBD COP decisions revealed that the clearest path for local community participation is via providing information directly to the CBD (either to the Executive Secretariat or other bodies, such as the Ad Hoc Working Group on Article 8(j)) and, to a lesser extent and more indirectly, through contributing to implementation.

Thus, the findings point to an area of local community involvement and influence that has not been covered to any great extent in the existing literature – influence sought through indirect routes that lie outside both intergovernmental negotiations and contentious social movement approaches. Various authors have pointed to the overlaps that have (re)developed between more organized civil society and social movements, which include some representatives of indigenous peoples and local communities – particularly around climate change issues after the disappointing Copenhagen COP of the UNFCCC (della Porta and Parks, 2014; Schlosberg and Collins, 2014). Yet a focus on how different local communities use 'everyday politics' to try and exert long-term change in global environmental governance appears to be missing from existing literature. The research presented here begins to fill that gap insofar as it makes the case for further research. This is not to argue that more visible action during negotiations or in movement protests are not necessary and important, rather that studies of these efforts can be completed by a focus on the everyday. Local communities do not easily fall into the category of either 'organized' civil society (NGOs) or social movements, but they are singled out as important actors in conservation, and research on how they can participate and shape global environmental governance is thus warranted. As Witter *et al.* (2015) point out with reference to the CBD – the focus of this study – 'progress is being made and leverage gained, but it is happening "slowly, slowly."' (908).

To elaborate on this research agenda, the following discussion situates the findings of the book in the literature on global civil society, which arguably includes this sort of everyday politics by non-state actors. This literature helps to locate local communities in the 'field' of global civil society along with NGOs and social movements. Then, the more critical literature on NGOization, illustrated briefly in Chapters 3 and 5, is discussed. This is because the role of NGOs as supporting actors for local communities and, potentially, in bringing local communities' views to the international level is suggested by the research presented here. In addition, the role of NGOs is also discussed in literature on democratic global governance (e.g. Slaughter, 2017). Yet there has been little explicit discussion of the critical view raised in the NGOization argument in work that theorizes a central role for NGOs in democratic governance. After a brief look at the literature that makes the normative case for local community participation in global environmental governance, the findings of the book are related to Dryzek and Stevenson's (2011) substantive model for democratic global governance of the environment, including reflections on NGOs.

While these observations move well beyond benefit-sharing, they are nonetheless based on the bottom-up study of local community framings related to it. Local communities placed benefit-sharing in a much broader view, and these discussions aim to begin to do the same.

Local communities within global civil society

A brief exploration of work on global civil society can help to situate local communities as actors within this broad field, as well as to reflect on the role of NGOs. Much of the literature on civil society tends to focus on NGOs alone, and NGOs that speak for public rather than private interests in particular. Yet historically the definition of civil society is broader, and international organizations tend to include much wider ranges of groups – including industry interests – in their definitions of civil society. Mary Kaldor's (2003a) work, where she develops the concept of global civil society, including her own normative view of the role it should play, is a useful source for mapping the development and breadth of what can be understood to constitute global civil society today.

Kaldor (2003a) identifies a number of historical definitions of civil society that have shifted over time, accompanying both political change and evolving social contracts between citizens and centers of authority. Earlier classical definitions of civil society, she notes, were more literal – the preoccupation was the creation of *a* civil society as opposed to an uncivil, conflict-ridden one. Expounded by classical European political theorists such as Hobbes and Locke, this definition was more closely linked to ideas about the rule of law and moving away from a so-called 'state of nature'.

Definitions of civil society then took on the view still shared by many today of a space between the private and public spheres linked to the market, though these views vary according to their correspondence with major ideologies (ibid.). Two more or less distinct versions emerge. The first resembles the De Tocquevillian view of 'associational life' where civil society is composed by economic interest groups such as guilds which are crucial for holding an ever-expanding State in check and to account. The second view is Marxist, and also places economic actors center stage. The balance of power is with civil society rather than the State in this view, as the State is understood as under the control of powerful market actors. In both views, civil society makes its wishes about political and market decisions known to State actors.

In more recent times these definitions of civil society have given way to what Kaldor (ibid.) dubs 'neoliberal' and 'activist' versions. Each builds on the two views already outlined. The neoliberal version accompanied the growth of the market alongside liberal citizenship regimes, and became more prominent after the end of the Cold War. Civil society here remains between the private sphere and the public, and provides services the market is not well suited to, forming a sort of safety net. The activist version of civil society challenges this rather apolitical view, and can be understood with reference to the neo-Marxist thought of Antonio Gramsci (ibid.). Gramsci refined the Marxist view of bourgeois civil society as an instrument of hegemony by looking at how this hegemony was exercised via lying in the discursive power of bourgeois civil society to establish norms and institutions. Civil society could challenge these by suggesting fresh norms and institutions, in particular through the education and media spheres. This view attributes a political role to civil society, seeing its roles as one of challenging dominant ideas where these do not serve citizens.

All of these historical views and definitions of civil society are tied to the State. Civil society exists to hold the State accountable, to fulfill some of its roles, or to challenge it on more fundamental levels. This symbiosis with the State creates a challenge for the creation of a *global* civil society, since no global State exists. Yet globally oriented social movements and NGOs and global economic actors are drivers and consequences of globalization processes, as are international organizations and agreements. Kaldor (2003a) thus explains the progressive emergence of this *global* civil society as a symbiotic relationship between these globally oriented civil society actors and international organizations that recalls the emergence of civil society within States.

Though international movements and NGOs emerged earlier, Kaldor (ibid.) dates the clear emergence of global civil society in waves of social movement activism, followed by waves of social movement 'taming', from the late 1960s onwards. The well-documented 1968 protest wave saw challenges mounted by groups of citizens across the globe, and translated into State concessions in the long term, even if these were not immediately apparent. Social movements in this view opened up spaces for change, as States realized that to stem further protest responses were needed. These accommodations led parts of social movements to institutionalize (or to be 'tamed' in Kaldor's (ibid.) terminology) in order to take advantage of more formal channels for exerting influence. Unlike earlier movements, the 1960s protest waves drew on universal ideas – they included themes of human autonomy (European student protests) and human rights (anti-Soviet protests, women's liberation, and others) in demands that focused on individuals but claimed a global reach.

This universalist trend and the pattern of protest and institutionalization repeated with waves in the mid-1970s and at the end of the Soviet Union. Movements for independence from colonial power and peace movements that contributed to the end of the Soviet Union also made global claims about autonomy, peace, human rights, and indeed the environment. These claims were made possible because of the new discursive spaces opened by previous protest waves, and the end of the Cold War presented an opportunity to push these universal claims at the global level in international organizations, newly keen and able to take on the language of the movements and open up a new global civil society. Unprecedented numbers of international NGOs were founded during this period (ibid.).

This *global* civil society also evolved in different ways which can be best understood with reference to the neoliberal and activist definitions introduced earlier. In the neoliberal view, global civil society became the preserve of international NGOs that now took over service provision in a global outlook in an era when the clash of ideologies had apparently been settled. This is the basis for the NGOization argument, which takes a critical view of the subsequent development of global civil society. The activist view, on the other hand, acknowledges the moment of institutionalization after the end of the Cold War, but notes that activist civil society – comprising both social movements and grassroots NGOs – became progressively more transnational in their challenges to a global market model from the late 1990s onwards via the global justice movement. Rather than echo this dichotomous view, Kaldor's definition sees both as part of the broad field of global civil society:

> a platform inhabited by activists (or post-Marxists), NGOs and neoliberals, as well as national and religious groups, where they argue about, campaign for (or against), negotiate about, or lobby for the arrangements that shape global developments. There is not one global civil society but many, affecting a range of issues [...]. It is not democratic; there are no processes of election [...]. It is also uneven and Northern-dominated. Nevertheless, the emergence of this phenomenon does offer a potential for individuals – a potential for emancipation.
> (Kaldor, 2003b: 590–591)

For the purposes of this discussion, it is this broad definition that helps to situate local communities in a broad field of global civil society. It highlights the heterogeneity within civil society and leaves space for a range of groups that can 'act globally' even if based locally. It also underlines that global civil society is not democratic in the liberal representative sense – a theme that will be returned to in the following section. Perhaps the most interesting part of the definition in view of democracy is the reference to the 'potential for emancipation' (ibid.: 591). In this view, the collective action of local communities, for example through the elaboration and use of community protocols or indeed other methods, is itself emancipatory to some extent – acting collectively can empower groups in and of itself. This chimes with more recent work on 'prefigurative' social movements in Europe and north America in particular (Kaldor, Selchow and Murray-Leach, 2015; Bailey *et al.*, 2017) where movement participants have sought to enact the democratic societies they advocate for on a global scale in miniature. This view imbues local community action with intrinsic worth beyond any eventual benefits brought about from relations with external actors, international participation, and the like.

Overall, the role of all the organizations and movements that populate global civil society is defined by Kaldor

> as the medium through which social contracts or bargains are negotiated between the individual and the centres of political and economic authority. Civil society is a process of management of society that is 'bottom-up' [...] It is about governance based on consent [...] In a global context, civil society offers a way of understanding the process of globalization.
>
> (Kaldor, 2003a: 142)

Global civil society in this view exists between private and public spheres, translating the needs of individuals into collective action to influence authorities at different levels. How global civil society – and local community action as part of it – can contribute concretely to 'governance based on consent' (ibid.) at the international level is discussed in more detail in the next subsection.

Another element in Kaldor's definition is the admission that global civil society is dominated by groups from the global North (ibid.). This leads to the critical view of global civil society, which sees the euro-centricity and ultimately colonial tendencies of the part of civil society defined as 'neoliberal' as damaging to activist civil society in the global South in particular. This is the core argument of 'NGOization' (Choudry and Kapoor, 2013), already sketched in brief in Chapters 3 and 5. This view sees international NGOs as dominating the field of global civil society to the detriment of community-based, grassroots groups and social movements.

In more detail, the NGOization argument sees neoliberal international NGOs as damaging for these groups as a result of a number of processes – 'the institutionalization, professionalization, depoliticization and demobilization of movements for social and environmental change' (Choudry and Kapoor, 2013: 1). The starting point of this argument refers to Kaldor's (2003a) neoliberal definition of civil society. It regards the majority of international NGOs as essentially complicit with State, market, and international institutions as a result of their dependence upon them. As these organizations come to dominate the global civil society field, spaces for other types of group shrink. This occurs through the aforementioned processes of NGOization. First, institutionalization refers to what Kaldor terms 'taming' (ibid.), that is when social movement waves are followed by openings in States and international organizations that lead parts of movements to reorganize as NGOs in order to take advantage of these. While Kaldor underlines discursive openings by authorities (ibid.), however, Choudry and Kapoor (2013) highlight that these are accompanied by increases in co-financing for NGO projects, particularly where these chime with the political will of the authorities concerned. At the end of the Cold War, for example, the authors note that co-financing was directed in particular towards NGOs delivering aid and development services that supported liberal market systems as part of a wider effort to democratize other States, for example through structural adjustment programs (ibid.; Kamat, 2004).

The second process of 'NGOization' is professionalization. This, Choudry and Kapoor (2013) argue, is when the dependence developed during institutionalization is translated into the structures of international NGOs. It can be read in organizational forms on one hand, and organizational logics on the other. As international NGOs rely on funds, or on their involvement in international organizations, their organizational forms shift to accommodate this. They follow legal frameworks that facilitate their involvement or role as project administrators by adopting more formal and hierarchical models. As numbers of professionalized NGOs increase, organizations also become more geared towards competing for access and funds, and are more preoccupied with survival than with their original goals. Professionally trained staff become the norm, and are increasingly university graduates from NGO management courses. This affects what comes to be accepted as the way that NGOs 'should' express themselves, and the knowledge this 'should' be based on. Paid staff take on the role of spokespeople for international NGOs, and begin to speak for other, more marginalized groups. This is described as the creation of a new elite, and of a 'form of hegemonic NGO politics' where professional NGOs function as gatekeepers between grassroots groups and movements on one hand, and States and international organizations on the other (ibid.).

The processes of institutionalization and professionalization lead in turn to the final two processes: depoliticization and demobilization. Depoliticization now spreads through international NGOs to affect local activist groups. As international NGOs transfer funds, they also transfer the obligations that have shaped them to the local level, either by directly choosing groups that do not challenge the political agendas of international organizations, or by indirectly encouraging local groups to reform in order to fit this bill. This fanning out of depoliticization to local levels thus ends in demobilization, also because those NGOs that are closer to authorities begin to denounce the 'wrong' kinds of groups (ibid.). Thus, large international NGOs that are closer to authorities are described as neo-colonial forces (ibid.) that contribute to the effort to export certain models of development by proposing 'development alternatives' rather than 'alternatives to development' (Escobar, 1995).

The overview of the story of global civil society emerges as the following: waves of social movements based on pushes for universal rights translate into waves of foundations of institutionalized NGOs. NGOs are seen in the critical literature as dominating the field of global civil society today, rather than forming one part of it. Scholars of environmental movements note a similar story. Dalton (1994) notes that the first wave of environmental movements focused on more limited questions of nature conservation. During the Cold War, a new wave of the environmental movement bridged issues of peace and anti-nuclear environmental activism, epitomized by Greenpeace. A third wave, following this logic, can be seen in the emergence of local groups and community activists addressing issues that link environmental suffering to inequality and racism (Shrader-Frechette, 2002). In more recent years, efforts to merge themes highlighted by organizations founded during the second wave of the environmental movement and the justice issues underlined by the third wave have been made, particularly as disappointment with the record of global environmental governance approaches via multilateralism has set in. This merging of issues is expressed in the environmental justice movement. The reflections that led to this new moniker are particularly useful for placing local community environmental activism within the field of global civil society. As already explored, the NGOization argument posits that international NGOs close to centers of authority dominate global civil society. Yet it also acknowledges some small shifts away from this trend (Choudry and Kapoor, 2013). The emergence of the climate justice movement is arguably a signal of one such shift as far as environmental questions are concerned.

To simplify a much more complex story (see Schlosberg, 2002), a moment of reflection among international NGOs coinciding more or less with the aftermath of the Copenhagen COP of the UNFCCC has been noted (della Porta and Parks, 2014; Schlosberg and Collins, 2014). This meeting was disappointing to many States who saw it as a failure of multilateralism, given the 'minilateral' solution brokered at the last minute. International NGOs were equally angry and disillusioned about the failure of high-level intergovernmental arenas to provide effective global approaches for environmental protection. This provoked a period of earnest debate in various organizations (della Porta and Parks, 2014). From this point onwards, conscious efforts were made to bridge the concerns of different social movements – about peace, social justice, and the environment – into a coherent whole by recognizing their interlinked nature within a frame of climate or environmental justice. Some international NGOs refocused their attention away from international organizations and towards 'partnership' approaches to fill gaps in global environmental governance in practical ways (see Stevenson, 2018, for an overview). Some of these approaches – for example the work of NGOs to support the communities presented in Chapter 2 – aim to empower communities to represent their positions and goals to a range of authorities.

This locates the importance of local community activism in the contemporary field of global civil society beyond the emancipatory dimension already mentioned. The NGOization argument makes the case for local collective action clear – without grassroots voices from local communities and social movements, international NGOs risk becoming progressively co-opted by international organizations and the States that compose them, though still seeking to follow their missions. When more NGOs focus on responding to the needs of local communities, and act to strengthen them, they can contribute to balance the field of global civil society alongside international NGOs, national NGOs, grassroots groups, social movements, and private interests. It is on the basis of this view that the next section reviews work about why the participation of local communities in global environmental governance is desirable, and how it could be achieved.

Local communities, participation, and democratic environmental governance

One argument in support of an increased role for local communities to participate in global environmental governance is derived from the NGOization view of an unbalanced field, as detailed in the previous section. Local community participation would help to balance the field, creating a more even spread between different types of civil society actors. A balanced field of global civil society actors can function better to represent the needs of different groups and individuals in society to centers of authority, including those at the international level (Kaldor, 2003a).

This view of global civil society hints that participation by a range of citizens is good for governance. In environmental governance, this is developed with the idea of civic environmentalism. Civic environmentalism sees stakeholder participation as a key ingredient for more effective and fair policies within a multi-level governance approach, and thus places local community participation in a pragmatic light. The reasoning behind the civic environmentalism (to borrow Sconfienza's (2017) terminology) argument is simple: participation will guarantee that policies are more fair and effective, as they will take account of more points of view held by a range of different stakeholders with varying interests. Thus, civic environmentalism sees participation as more than simply what is just for stakeholders affected by a decision, and suggests that it is also necessary to make decisions effective. Bäckstrand

and Lövbrand (2006, 2016) place civic environmentalism within a longer history of the global environmental governance regime on climate change in particular (see also Dryzek, 2005). Civic environmentalism succeeds and builds upon preceding ideas of 'green governmentality' and 'ecological modernization' (Bäckstrand and Lövbrand, 2016). 'Green governmentality' focuses on ecological limits and the need for the rational management of environmental problems through a professional bureaucracy. Ecological modernization builds on this, but recognizes that because environmental problems are extremely complex and involve various areas of decision-making, decentralized approaches are needed to solve them. Civic environmentalism, finally, makes the implicit participatory element of ecological modernization explicit. Environmental questions are complex sets of problems that can best be solved in a multi-level governance approach and with the inputs of a wide range of governmental and non-governmental actors.

Civic environmentalism thus echoes the view of global civil society proposed by Kaldor (2003a). In addition, Bäckstrand and Lövbrand (2016) link the emergence of this version of civic environmentalism to the UNFCCC Copenhagen COP discussed above. The reflective moment within global civil society, where a range of organizations considered their roles and how to best fulfill their aims, was therefore matched by a similar moment in institutions of global governance. This resulted in a turn towards so called partnership approaches to provide practical solutions to environmental issues in the absence of an effective intergovernmental approach. Talk about participation in the UNFCCC as a result of the civic environmentalist turn can thus be argued to be at least partly a response to criticism from different actors in global civil society during and after Copenhagen. References to participation retain the positive elements of green governmentality and ecological modernization, but add an element of fairness that responds to the arguments of the environmental justice movement (though more radical movement groups disagree that participation is enough to solve the problems of global environmental governance).

In the CBD, the main focus in this book, the peak in talk about participation is at COP 7 in 2004 (following the data presented in Chapters 4 and 5), some years before the turn in the UNFCCC. A possible explanation for this is the view of the CBD as a frontrunner in terms of actions for the inclusion of indigenous peoples and local communities – as mentioned elsewhere, the CBD has been described as the arena looked to by others for best practices on the involvement of non-state actors (Affolder, 2017; Schabus, 2017). The CBD's turn towards participation is addressed more specifically to indigenous peoples and local communities, but can conceivably be seen as another contributing factor in the UNFCCC's more general turn towards participation. If this turn is understood as a response to arguments about environmental justice, this view makes more sense – indigenous peoples and marginalized local communities are central to such arguments (Shrader-Frechette, 2002).

The arguments presented thus far about why local community participation in global environmental governance is desirable speak more to fairness (NGOization arguments and the need for balances between actors in global civil society) and efficiency (civic environmentalism). A broader normative argument for participation is suggested in Kaldor's definition of global civil society. This definition refers to social contracts, bottom-up management, and governance by consent in a global context (Kaldor, 2003a: 142). Arguments about environmental justice also underline the normative reasons for participatory environmental governance as a form of substantive democracy (Shrader-Frechette, 2002; Martin, 2017). The literature on deliberative democratic approaches to environmental governance sheds light on *how* participation might be achieved, and how it can equate with democratic global environmental governance.

The big idea of deliberative democracy is to rethink how collective decisions are made. In representative democracies, interests are considered in aggregate and the preference of the majority is generally taken as the basis for decisions that affect all members of the polity in question, including the minorities that do not share that preference. Deliberative democratic theory suggests a different method, where consensus is reached through informed debates among equals. Deliberative democratic theory thus considers the true collective interest as a consensus, not as the sum of individual opinions. If decisions are based on wide-ranging input and consensus is reached, the legitimacy of the resulting decisions also increases. Deliberation is seen as democratic – and its decisions as legitimate – because it implies more than authorities merely listening to various points of view before taking a decision. Deliberation instead involves rational debate geared towards consensus over decisions – decision-making is located in the debate rather than after it. Deliberation is one way to build a mechanism for bottom-up governance into democracy.

The idea of deliberative democracy is generally attributed to Jürgen Habermas' work on the public sphere, ideal speech situations, and communicative action (Habermas, 1984, 1987). In an ideal speech situation, Habermas suggests, citizens can discuss with one another freely, unfettered by distortions created by unequal power relations. Those involved will eventually come to a decision that may not match the initial preference of any one person or group of people – rather the outcome will reflect a consensus decision based on the exchange of opinion and information. Decisions taken in this manner should be more conducive to satisfying social justice concerns, as they should see participants convinced about which decision is the best for the majority, or for the most in need, or other criteria decided in debate. Decisions should be widely accepted, because even where the consensus does not reflect an individual's ideal outcome, they have been rationally persuaded to back it. In other words, deliberation gives a better chance of changing a participant's mind.

This theoretical scenario of deliberation is an ideal type. Real-life experiments with deliberative democracy, as well as other theoretical work, have highlighted that achieving a completely equal, undistorted forum for deliberation is nigh on impossible (Young, 2004). The presence of power in exchanges between citizens cannot be entirely extinguished given the deep-set nature of some forms of power and structural barriers to certain forms of expression (ibid.). In local communities, for example, power (or the lack of it) may be attached to language, social status, age, gender, or other characteristics. These can trigger assumptions about speakers, including about their perceived right to speak, which can affect the weight given to their opinions. In the ideal speech situation, deliberation must take place between individuals who hold no assumptions about one another for unbiased rational decisions to be taken. In reality, unconscious biases exist. Similar assumptions may also apply to participants' judgements of any experts called on to present information to participants in a deliberative forum. The expert may not be perceived as trustworthy: for example, drawing on the Argentina case study presented in Chapter 2, if an industry representative provides information about a project her or his employers wish to pursue, questions about the impartiality of the evidence are likely to arise. Power, in other words, is also expressed through information, or the selective (or false) presentation of information. Despite this, deliberation as a mechanism for bottom-up and more substantively democratic decision-making remains a good fit for local community participation in global environmental governance.

This is suggested by a number of the findings in this book. First, deliberative democracy resonates with the common themes that emerged from the case study comparison (Chapter 3). These revolved around central questions of recognition and respect for different worldviews, and pointed to the importance of meaningful dialogue between communities and external actors, as well as within communities themselves. Deliberation answers this emphasis on meaningful and equal communication and places it in a view of substantive democratic decision-making. The advantage of a deliberative democratic view for global environmental governance is that it can respond to the needs of diverse local communities in a manner that representative mechanisms cannot. Indeed, various scholars have underlined that representative democracy is rooted in western philosophical traditions that do not answer to the more substantive conceptions of democracy that developed elsewhere in the world (Murphree, 2004; Sen, 1999) and that representative models are not conducive to effective environmental governance at the local level (Child and Barnes, 2010).

Deliberative democratic ideas also echo the core themes uncovered in CBD decisions relevant to indigenous peoples and local communities: recognition (once again) and participation. Recognition for the CBD refers to traditional knowledge and customary sustainable use by local communities and indigenous peoples, yet by extension refers to both recognition in the sense given in the case comparison, and the recognition of interlocutors in deliberations. For participation the link is clear, and relates to inputs from a wide range of actors as seen in the idea of civic environmentalism. This is imbued with democratic potential through the lens of deliberative democracy, which describes participation beyond the provision of 'inputs' – one of the clearest paths for participation discussed in Chapter 5. Participation in deliberative democracy is rational debate on the basis of different inputs geared towards a better understanding of all views to ultimately arrive at a shared consensus decision. This is close to the vision of meaningful dialogue sought by the local communities studied in this book.

Finally, ideas of deliberative democracy also chime with the discussion of community protocols as potential tools for local communities to communicate their points of view to authorities, including at the international level. In Chapter 5 this discussion pointed to the importance of participatory, inclusive processes within communities to arrive at internal consensus around the content of protocols. This corresponds with the idea of deliberation, and can be reframed as underlining the importance of local deliberative democracy to feed into other decision-making arenas. This opens up a first idea about what deliberative global environmental governance could look like insofar as the role of local communities is concerned. Several points support this. Deliberative democratic theory essentially places deliberation in the position of the community in republican democratic theory. The community – those in a polity holding political rights – is the repository of popular sovereignty in republican models. In deliberative democracy, deliberation itself is the repository of popular sovereignty (Rittberger, 2010). Deliberation within the local community thus takes on a democratic function in this model, and indeed takes on the role of a *community-building process* – an idea discussed in Chapter 3. This also recalls the idea of emancipation in Kaldor's (2003b) definition of global civil society. By deliberating, the community defines and constitutes itself and through this comes emancipation. The deliberations of the community then feed into wider democratic governance processes.

In this view the number of studies that end by advocating for some kind of deliberative democracy for global governance appears justified. Kaldor's (2003a) work ends with recommendations for deliberation, as does the work of many scholars of social movements. Work on social movements also underlines the practicality of deliberation: deliberative experiments in social movements have come the closest to ideal speech situations by addressing power imbalances in explicit ways (Castells, 2012; della Porta and Rucht, 2013). This underlines the potential of community protocols – and other community-based collective action processes – in models of global environmental governance. Dryzek (2005) also concludes his study of the discourses of environmental politics by underlining that deliberation is a promising avenue for combining different views – each of which proposes different solutions according to different problem definitions – in the most effective and elastic ways to address questions of environmental protection. Radical critique from social movements, he adds, is necessary to train the attention of established authorities on alternative understandings of old and new problems. Local communities can be added to this as another group of actors in global civil society taking part in deliberation.

This brief and non-exhaustive overview of deliberative democracy, the points of convergence with the findings of this book, and its recurring popularity in accounts of bottom-up participation suggest it is a fruitful area for further study. Where this book differs to some extent from other accounts of the potential of deliberative democracy in global governance is in its departure point – local communities. While accounts of how deliberative processes might be worked into global governance assume that deliberation is desired by citizens, including local communities, this book has arrived at a conclusion about the importance of deliberative processes on the basis of findings from local community case studies. Nevertheless, calls for international deliberation did not emerge in the local community cases here. The interest in the international level flowed instead from the international source of the policy area of interest – benefit-sharing. In addition, scholarship shows that local involvement in environmental governance gives better results, but that top-down applications of international environmental policies are often problematic at local levels. The interest in how local democratic deliberation may feed into international chains of participation follows this logic. To conclude these reflections, the next subsection takes a brief look at literature on practical models of global environmental governance and how local deliberation could fit into these. The aim of this discussion is to suggest some potential features of a locally sensitive global regime.

Local communities, global processes: potential features of deliberative global environmental governance

A key question for scholars of deliberative democracy is how to achieve it in practice. Can deliberative models be 'scaled up' to national and international levels? The answer for most scholars is that, in practice, deliberation should supplement rather than entirely replace liberal representative models. Deliberation within defined groups will feed into existing decision-making forums. These defined groups may be parliaments (Steiner *et al.*, 2004), social movements (della Porta and Rucht, 2013; Slaughter, 2017), or civil society more generally (Kaldor, 2003a). The key point for 'scaling up' deliberative democratic forms of decision-making identified in the work of Mansbridge *et al.* (2012) is that not every part of a deliberative system need be deliberative in itself (see also Dryzek and Stevenson, 2011).

The most developed practical view of a deliberative system of global environmental governance is the work by Hayley Stevenson and John S. Dryzek (Dryzek and Stevenson, 2011; Stevenson and Dryzek, 2014). These authors depart from the problem of liberal democracy for environmental politics in particular – including the slow pace of decision-making for urgent problems and the western roots of liberal representative democracy. Given the fairness, legitimacy, and efficacy returns from democratic environmental governance already discussed above, they seek to outline a model of global environmental deliberative decision-making that combines both deliberation, effective decision-making, and radical critique for innovation (a need identified in Dryzek's earlier (2005) work).

They first identify various qualities of deliberative systems, before evaluating the potential for such qualities to exist in the climate change regime. A deliberative system, they argue, must include: 1) a public space where a range of actors with different views interact freely; 2) an empowered space for decision-making; 3) transmission, whereby the public space influences the empowered space; 4) accountability between the empowered space and the public space; 5) meta-deliberation about the system as a whole leaving space for radical change if needed; and, 6) decisiveness, meaning that deliberation must count towards final decisions (paraphrasing Dryzek and Stevenson, 2011: 1866–1867). Although their findings with regard to the potential for deliberation in the climate change regime are mixed, they conclude that the seeds for building deliberation are present (ibid.).

The question of local community participation under discussion here concerns the transmission aspect of Dryzek and Stevenson's (2011; 2014) model – that is the way that actors in the public space transmit ideas and influence actors in the empowered space. While the authors observe the ways in which different types of civil society actors function to transmit ideas, they do not focus on whether these actors are deliberative or not, or, where civil society actors are concerned, on the critical arguments of NGOization. In this vein, Pickering (2019) suggests that further analysis is needed to inspect these 'deliberative ecologies'. The discussions above about why local communities are needed to balance the presence of more formal international NGOs in global civil society answer this call, as do the discussions about deliberation in the development of community protocols. The idea of transmission is thus a useful one to illustrate how the findings of this book fit into and expand Dryzek and Stevenson's model of deliberative global environmental democracy (Dryzek and Stevenson, 2011; Stevenson and Dryzek, 2014). Their comments about insider and outsider actors who move between the public and the empowered spaces and back provide a basis to speculate about the roles of local communities and NGOs in a deliberative model of global environmental governance.

Two images are suggested in this vein by the findings of this book. One is of local communities acting in global civil society independently, using the different available paths for participation in the CBD to bring their views – decided in deliberative processes in local settings – to that arena and thus improve and diversify the transmission between the public and the empowered spaces. This, as discussed, could take place through the provision of inputs, implementation, or indeed the types of action taken at COP meetings analyzed in other literature (Marion Suiseeya, 2014; Witter *et al.*, 2015; Eastwood, 2019; Marion Suiseeya and Zanotti, 2019). However, this assumes that local communities desire and have the capacity to participate in international arenas – a fact that does not, as mentioned, emerge from the (limited) number of case studies presented. A second image, which finds more confirmation in both the existing literature and the case studies, sees local communities contribute to transmission between the public and the empowered spaces through local deliberative action which is supported and subsequently communicated by NGOs with the will and capacity to act in international arenas and to exploit available paths for participation. In this view, the importance of the type of NGO involved comes to the fore in light of arguments about NGOization processes.

NGOs that can act to improve this type of transmission and feed into deliberative global environmental governance alongside local communities, as well as support deliberation in local communities (and their eventual empowerment to act independently in deliberative global environmental governance should they so wish), need to avoid the NGOization processes identified in critical literature (Choudry and Kapoor, 2013) and contribute to re-balance the field of global civil society. This is not easy to accomplish – it requires both long-term commitment to local community work and familiarity and proximity to arenas of global environmental governance that are challenging to achieve, particularly given the funding structures and organizational requirements that exist at present. Further research looking more closely at the composition of the public space and deliberation at the micro level, and networks among local communities and different types of NGOs at the meso level, would shed further light on this possibility. Nevertheless, existing findings about the achievement of deliberative democracy in social movements, combined with the moment of reflection identified in the environmental movement, suggest that, with the right support from institutions of global environmental governance committed to moving towards more democratic models, it is possible.

Overall, this view suggests there is a potential important role for local communities to contribute to a fairer and more democratically legitimate version of global environmental governance. Ultimately, such a model would be better placed to design and implement policies of fair and equitable benefit-sharing which seem unlikely to succeed on the basis of current approaches. The modest contribution of this book has been mostly empirical – it adds comparative case studies and a dynamic and detailed analysis of discursive spaces for local issues in the CBD. Yet it has also sought to underline the need to consider such questions from the bottom up, and to pay attention to the potential and importance of everyday politics for global questions. Policies such as fair and equitable benefit-sharing will only work, this book suggests, in the context of a more democratic model of global environmental governance that allows for locally sensitive policies to emerge from deliberative processes.

Note

1 Chapter 4 also shows a dip in discourses about recognition at COP 10, though this is based on analysis of all the decisions taken at that meeting, and is not the lowest dip over time.

References

Affolder, N. (2017) 'Non-State Actors', in Morgera, E. and Razzaque, J. (eds) *Encyclopedia of Environmental Law: Biodiversity and Nature Protection Law*. Cheltenham: Edward Elgar, pp. 387–398.

Bäckstrand, K. and Lövbrand, E. (2006) 'Planting trees to mitigate climate change: contested discourses of ecological modernization, green governmentality and civic environmentalism', *Global Environmental Politics*, pp. 50–75. doi: 10.1162/glep.2006.6.1.50.

Bäckstrand, K. and Lövbrand, E. (2016) 'The road to Paris: contending climate governance discourses in the post-Copenhagen era', *Journal of Environmental Policy and Planning*, pp. 1–19. doi: 10.1080/1523908X.2016.1150777.

Bailey, D. J. *et al.* (2017) *Beyond Defeat and Austerity. Disrupting (the Critical Political Economy of) Neoliberal Europe*. London: Routledge.

Bavikatte, K. (2014) *Stewarding The Earth. Rethinking Property and the Emergence of Biocultural Rights*. Oxford: Oxford University Press.

Bavikatte, K., Robinson, D. F. and Oliva, M. J. (2015) 'Biocultural community protocols: dialogues on the space within', *IK: Other Ways of Knowing*, 1(2), pp. 1–31.

Brand, U. and Vadrot, A. B. M. (2013) 'Epistemic selectivities and the valorisation of nature: the cases of the Nagoya Protocol and the intergovernmental science-policy platform of the biodiversity and ecosystem services (IPBES)', *Law, Environment and Development Journal*, 9(2), pp. 202–220.

Castells, M. (2012) *Networks of Outrage and Hope: Social Movements in the Internet Age*. Cambridge, UK: Polity Press.

Child, B. and Barnes, G. (2010) 'The conceptual evolution and practice of community-based natural resource management in southern Africa: past, present and future', *Environmental Conservation*, 37(3), pp. 283–295.

Choudry, A. and Kapoor, D. (2013) *NGOization: Complicity, Contradictions and Prospects*. London: Zed Books.

Convention on Biological Diversity (2004a) *Addis Ababa Principles and Guidelines on the Sustainable Use of Biodiversity*. CBD Decision VII/12, Annex, Montreal.

Convention on Biological Diversity (2004b) *Akwé: Kon Voluntary Guidelines for the Conduct of Cultural, Environmental and Social Impact Assessment Regarding Developments Proposed to Take Place on, or which are Likely to Impact on, Sacred Sites and on Lands and Waters traditionally Occupied or Used by Indigenous and Local Communities*. CBD Decision VII/16, Annex, Montreal.

Dalton, R. J. (1994) *The Green Rainbow. Environmental Groups in Western Europe*. New Haven, CT: Yale University Press.

Dryzek, J. S. (2005) *The Politics of the Earth. Environmental Discourses*. Second edition. Oxford: Oxford University Press.

Dryzek, J. S. and Stevenson, H. (2011) 'Global democracy and earth system governance', *Ecological Economics*, 70(11), pp. 1865–1874. doi: 10.1016/j.ecolecon.2011.01.021.

Eastwood, L. E. (2019) *Negotiating the Environment. Civil Society, Globalisation and the UN*. London: Routledge.

Escobar, A. (1995) *Encountering Development: The Making and Unmaking of the Third World*. Princeton, NJ: Princeton University Press.

Garrido, M. and Halavais, A. (2003) 'Mapping Networks of Support for the Zapatista Movement: Applying Social Networks Analysis to Study Contemporary

Social Movements', in McCaughey, M. and Ayers, M. D. (eds) *Online Activism in Theory and Practice*. London: Routledge, pp. 165–184.

Habermas, J. (1984) *The Theory of Communicative Action, Vol. 1. Reason and the Rationalization of Society*. Boston: Beacon Press.

Habermas, J. (1987) *The Theory of Communicative Action, Vol. 2. Lifeworld and System*. Boston: Beacon Press.

Kaldor, M. (2003a) *Global Civil Society: An Answer to War*. Cambridge, UK: Polity Press.

Kaldor, M. (2003b) 'The idea of global civil society', *International Affairs*, 79(3), pp. 583–593.

Kaldor, M., Selchow, S. and Murray-Leach, T. (eds) (2015) *Subterranean Politics in Europe*. Basingstoke: Palgrave Macmillan.

Kamat, S. (2004) 'The privatization of public interest: theorizing NGO discourse in a neoliberal era', *Review of International Political Economy*, 11(1), pp. 155–176.

Keck, M. E. and Sikkink, K. (1998) *Advocacy Networks in International Politics*. Ithaca, NY: Cornell University Press.

Mansbridge, J. *et al.* (2012) 'A Systemic Approach to Deliberative Democracy', in Parkinson, J. and Mansbridge, J. (eds) *Deliberative Systems: Deliberative Democracy at the Large Scale*. Oxford: Oxford University Press, pp. 1–26.

Marchegiani, P., Morgera, E. and Parks, L. (2019) *Indigenous peoples' rights to natural resources in Argentina, the challenges of impact assessment, consent and fair and equitable benefit-sharing in cases of lithium mining*, BENELEX Working Papers, 19. Available at: https://papers.ssrn.com/sol3/papers.cfm?abstract_id=3317375.

Marion Suiseeya, K. R. (2014) 'Negotiating the Nagoya Protocol: indigenous demands for justice', *Global Environmental Politics*, 14(3), pp. 102–124.

Marion Suiseeya, K. R. and Zanotti, L. (2019) 'Making influence visible: innovating ethnography at the Paris Climate Summit', *Global Environmental Politics*, 19(2), pp. 38–60.

Martin, A. (2017) *Just Conservation. Biodiversity, Wellbeing and Sustainability*. London: Routledge.

Morgera, E. (2017) *Reflections on 2016 UN Biodiversity Conference (Parts I & II): assessing the Mo'otz kuxtal guidelines on benefit-sharing from the use of traditional knowledge*, BENELEX Blog.

Morgera, E. (2018) 'Dawn of a new day? The evolving relationship between the Convention on Biological Diversity and international human rights law', *Wake Forest Law Review*, 54, pp. 101–121.

Murphree, M. (2004) 'Communal approaches to natural resource management in Africa: from whence to where?', *Journal of International Wildlife Law and Policy*, 7, pp. 203–216.

Parks, L. (2018) 'Spaces for local voices? A discourse analysis of the decisions of the Convention on Biological Diversity', *Journal of Human Rights and the Environment*, 9(2), pp. 141–170.

Pickering, J. (2019) 'Deliberative ecologies: complexity and social-ecological dynamics in international environmental negotiations', *Global Environmental Politics*, 19(2), pp. 61–80.

della Porta, D. and Parks, L. (2014) 'Framing Processes in the Climate Movement: from Climate Change to Climate Justice', in Dietz, M. and Garrelts, H. (eds) *Routledge Handbook of Climate Change Movements*. London: Routledge, pp. 19–30.

della Porta, D. and Rucht, D. (eds) (2013) *Meeting Democracy*. Cambridge, UK: Cambridge University Press. doi: 10.1017/CBO9781139236034.

Reimerson, E. (2013) 'Between nature and culture: exploring space for indigenous agency in the Convention on Biological Diversity', *Environmental Politics*, 22(6), pp. 992–1009.

Ribot, J. C. and Peluso, N. L. (2003) 'A theory of access', *Rural Sociology*, 68(2), pp. 153–181.

Rittberger, B. (2010) 'Democracy and European Union Governance', in Egan, M., Nugent, N., and Paterson, W. E. (eds) *Research Agendas in EU Studies: Stalking the Elephant*. Basingstoke: Palgrave Macmillan, pp. 134–167.

Robinson, D. F. (2015) *Biodiversity, Access and Benefit-Sharing. Global Case Studies*. London: Routledge.

Schabus, N. (2017) 'Traditional Knowledge', in Morgera, E. and Razzaque, J. (eds) *Encyclopedia of Environmental Law: Biodiversity and Nature Protection Law*. Cheltenham: Edward Elgar, pp. 264–279.

Schlosberg, D. (2002) *Environmental Justice and the New Pluralism: The Challenge of Difference for Environmentalism*. Oxford: Oxford University Press.

Schlosberg, D. and Collins, L. B. (2014) 'From environmental to climate justice: climate change and the discourse of environmental justice', *WIREs: Climate Change*, 5(3), pp. 359–374.

Sconfienza, U. (2017) *The politics of environmental narratives*. Tilburg University.

Sen, A. K. (1999) 'Democracy as a universal value', *Journal of Democracy*, 3(3), pp. 3–17.

Shrader-Frechette, K. (2002) *Environmental Justice: Creating Equity, Reclaiming Democracy*. Oxford: Oxford University Press.

Slaughter, S. (2017) 'The G20 and global justice: the potential of transnational deliberative democratic theory', *Global Society*, 31(4), pp. 460–478. doi: 10.1080/13600826.2017.1325844.

Steiner, J. *et al*. (2004) *Deliberative Politics in Action: Analysing Parliamentary Discourse*. Cambridge, UK: Cambridge University Press.

Stevenson, H. (2018) *Global Environmental Politics. Problems, Policy and Practice*. Cambridge, UK: Cambridge University Press.

Stevenson, H. and Dryzek, J. S. (2014) *Democratizing Global Climate Governance*. Cambridge, UK: Cambridge University Press.

Swemmer, L. Grant, R., Annecke, W. and Freitag-Ronaldson, S. (2015) 'Toward more effective benefit-sharing in South African national parks', *Society & Natural Resources*, 28(1), pp. 4–20.

Tysiachniouk, M. and Olimpieva, I. (2019) 'Caught between traditional ways of life and economic development: interactions between Indigenous peoples and an oil company in Numto Nature Park', *Arctic Review of Law and Politics*, 10, pp. 56–78.

Witter, R. *et al*. (2015) 'Moments of influence in global environmental governance', *Environmental Politics*, 24(6), pp. 894–912.

Wynberg, R. and Hauck, M. (eds) (2014) *Sharing Benefits from the Coast. Rights, Resources and Livelihoods*. Cape Town: University of Cape Town Press.

Young, I. M. (2004) *Inclusion and Democracy*. Oxford: Oxford University Press.

Index

Note: Bold page numbers refer to tables; *italic* page numbers refer to figures and page numbers followed by "n" denote endnotes.

For Product Safety Concerns and Information please contact our EU
representative GPSR@taylorandfrancis.com
Taylor & Francis Verlag GmbH, Kaufingerstraße 24, 80331 München, Germany

www.ingramcontent.com/pod-product-compliance
Lightning Source LLC
LaVergne TN
LVHW011129300926
849160LV00008B/581
* 9 7 8 1 0 3 2 0 8 3 9 1 9 *